How To Make the OSHA—1970 Work For You

How To Make the OSHA—1970 Work For You

Handbook of the Williams-Steiger
Occupational Safety and Health Administration

Includes New 1976 OSHA-NIOSH Personnel Directories

DAVID R. SHOWALTER
President
Human Resources Conservancy
Tacoma, Washington

ANN ARBOR SCIENCE
PUBLISHERS INC
P.O. BOX 1425 • ANN ARBOR, MICH. 48106

Second Printing, 1974

Third Printing, 1976

© copyright 1972 by Ann Arbor Science Publishers, Inc.
P.O. Box 1425, Ann Arbor, Michigan 48106

Library of Congress Catalog Card No. 72-88890
ISBN 0-250-40006-5

All Rights Reserved

Printed in the United States of America

CONTENTS

CHAPTER 1. INTRODUCTION 1

CHAPTER 2. WHO'S ON FIRST? 3

**CHAPTER 3. READTHROUGH OF THE ACT—
            SALIENT POINTS** 9
   Eight Requirements for Acceptance of State Plan 22
   Option to Enforce Federal Standards 22
   State Plans to be Evaluated 23
   Contents of 18h Agreement (State's Application) 23

CHAPTER 4. SAFETY COORDINATOR/DIRECTOR .. 25
   Selection 25
   The Safety Director 28

CHAPTER 5. SAFETY AND HEALTH COMMITTEE .. 31

CHAPTER 6. SAFETY LOG 35
   Letter of Intent 37

**CHAPTER 7. NOISE ABATEMENT AND HEARING
            CONSERVATION PROGRAM** 39
   Occupational Noise Standards 39

Audiometry 41
    Records 42
    Selection of Personal Protective Equipment 42
    Hearing Conservation Program 42
    Compliance Plan 43
Comments on Hearing Conservation 44
Letter to Employee Reporting Hearing Test Results 48
Annual Audiogram Form Letters 48

CHAPTER 8. COMMUNICATION AND EDUCATION 51

Information Dissemination 51
Training and Education 54
Public Relations—Information 54
Motivational and Communicating Techniques
    (How to Get Job Done) 55
Employee Relations 58

CHAPTER 9. INDEX TO PART 1910 63

Occupational Safety and Health Standards,
    Federal Register Vol. 36, Number 105 63

CHAPTER 10. TARGET HEALTH HAZARDS PROGRAM 83

Asbestos 83
Carbon Monoxide 84
Cotton Dust 85
Lead 87
Silica 88

CHAPTER 11. NIOSH 91

Functional Organization 91
HEW Responsibilities 92
Toxicity Rulings 93
Educational Programs 94

Administrative Responsibility 95
    Office of the Director 95
    Office of Public Information 96
    Office of Extramural Activities 96
    Office of Administrative Management 96
    Office of Planning and Resource Management 96
    Office of Research and Standards Development ... 96
    Office of Manpower Development 97
    Office of Health Surveillance and Biometrics 97
    Division of Laboratories and Criteria Development .. 97
    Division of Field Studies and Clinical
        Investigators 97
    Division of Technical Services 98
    Division of Occupational Health Programs 98
    Division of Training 98
    Appalachian Laboratory for Occupational
        Respiratory Diseases 98
    1972 Priority List for Criteria for Toxic Substances
        and Physical Agents 98
    National Institute for Occupational Safety and Health
        Priority List for Criteria for Toxic Substances
        and Physical Agents, 1972 99
    National Surveillance Network 101
    National Occupational Hazard Survey Form 101

CHAPTER 12. CONCLUSION 107

CHAPTER 13. SAFETY INSPECTION GUIDE 111
How to Make a Safety Inspection 111
    Detecting Unsafe Practices, Conditions
        and Hazards 111
Main Points to Look For 113
    Unsafe Conditions 113
    Unsafe Practices 113
Reference Section 114

vii

CHAPTER 14. NATIONAL PERSONNEL ADMINISTRATING THE OCCUPATIONAL SAFETY AND HEALTH ACT OF 1970 127

Department of Labor 127
Occupational Safety and Health Administration Regional, Area and District Offices; Names of Coordinating Personnel 128
Bureau of Labor Statistics Regional Offices 143
National Institute for Occupational Safety and Health 144
Private Associations Concerned with Safety and Health 146

INDEX 147

CHAPTER ONE
Introduction

This book is intended for both management and employee. It has direct application to president, union member and worker alike, and it can be immeasurably helpful in meeting the requirements of an Act (Federal and state) which, necessary though it may be, still remains maligned, criticized and misunderstood by management, labor, the safety professional (who ought to know better) and even the worker for whom it was intended.

Few people seem to recognize the potential (let alone the merit) of its basic employee relations aspect. If such recognition is granted it is accorded a negative rather than positive emphasis.

Many people still argue about its wordage. The Act, in plain truth, is about safety, not semantics. It seeks to establish a safe and healthful workplace for all workers.

Although not stated in the following way, the objective of the Williams-Steiger bill is fourfold.

(1) to reduce the number of fatalities occurring annually in the American workplace
(2) to reduce the number, severity and cost of job-related injuries and health problems
(3) to help establish the security and well-being of those who do the job for us
(4) to reduce profit-loss and increase profit-potential for business and industry.

That is an intelligent and practical assessment of what the Act really is and is a 4-point objective rather difficult, impractical, if not indeed downright unintelligent, to argue with.

The validity of the fourth point is substantiated by the insurance industry again and again, year after year. Accidents are phenomenally costly and are paid for from the profit side of the ledger—not the production side. As a matter of simple fact it is more accurate to say that in the end it is the consumer who really pays for industrial fatalities and injuries.

Assistant Secretary of Labor, George C. Guenther, the first federal department head to be charged with the new responsibility contained in the Act, stated, that *protectivity*—a new concept in the safe and healthful work place laws—makes the gains of *productivity* possible. The point is well made. Whether or not "protectivity" becomes equal to "productivity" as an integral and essential part of profit-making for American business and industry remains to be seen.

The truth, nonetheless, is incontestable. With few exceptions, profit-making has never yet considered loss control, risk management, accident prevention, safety—however, it is termed—as anything other than part of production and of negative, minimum, or no importance if it interfered in any way with maximum production.

What the Assistant Secretary of Labor was suggesting and what the Act may bring into sharp focus is that safety as such is a separate and important issue; it is a part of the profit-making or profit-losing process; it is of equal importance with production, sales and merchandising, advertising and promotion, public and community relations, and when successfully managed helps make the whole process that much more profitable.

CHAPTER TWO
Who's on First?

As president, owner, executive officer, by whatever title (if you are the individual where the buck stops passing) you will need to be involved with the Act (Federal and state), at least to the point where you know what it is, how it applies to you, and generally what needs to be done about it. It is embarrassing, not to mention impractical, unprofitable and unintelligent, to find your company on the short end of the "inspection stick," cited with several alleged violations and fined hundreds or thousands of dollars because somebody didn't get the message.

It has happened. As a very busy, involved and active chief executive of a company faced with all the problems and details that successful management requires, you would find the above circumstance very irritating indeed.

The alleged violations, however, had their origin with you. To avoid such a situation, you must first read the Act. Next you need to assign the full responsibility for it to some individual. He should then always report directly to you. Then you need to act. Do something about implementing the Act. Some suggestions are made in the following pages.

Don't think, or let anyone else think, that the Act will just go away or that it doesn't apply to you. It won't go away. The enactment of a law by the Congress of these United States of America takes some doing in the first place. Once enacted it doesn't rescind or even amend too easily.

Similar enabling Acts by 50 state legislatures, providing states with the administration of their own safety and health regulatory procedures, bring it even closer to home on a local functioning level. This further emphasizes the fact that the Act(s) is/are here to stay. And, while nobody can deny the ameliorating effect that political or vested interest pressures exert on any legislation, there is, in this case, a very capable, interested and vocal proponent and guardian of it—organized labor.

The Act applies to you if you have one or more employees working for you.

It has a more intensified, more comprehensive application if your business falls within any designation of *target industry*—those industries with the highest accident frequency rates. The first of these are longshoring, meat and meat packing, wood products, sheet metal and roofing, and mobile home construction.

One of OSHA's first approaches within the federal administration of the Act was to inspect these five target industries. What the next five will be is anybody's guess, but a look at the U. S. Department of Labor's statistical analysis of accidents by industry is a good indication. A "Target Health Hazards" program was established focusing on five toxic substances which affect workplaces—asbestos, cotton dust, silica, lead and carbon monoxide. OSHA inspection priorities included both the new program and industrial hygiene inspections carried out in thousands of workplaces.

In the event of a job-related fatality or an accident hospitalizing five or more workers, a CSHO (Compliance Safety and Health Officer) is on the spot to make an investigation.

The program also provides for an almost immediate investigation of an employee request for an inspection of a workplace considered to be unsafe or unhealthful, provided that the "request" is stated with "reasonable particularity," and is basically valid. The possibility that it may be the object of an inspection by virtue of the "random sampling" part of the inspection program, used along with the "target industry" approach, can not be overlooked by any business.

The Act (Federal and state) in its voluntary compliance principle, operates similarly to IRS. There are not now, nor will there ever be, enough compliance officers to annually inspect all business establishments affected by the Act. However, the possibility, even the probability, of inspection always exists. The fines and penalties attached to failure to comply with the standards are a healthy deterrent factor.

Another deterrent, although not given much recognition at this point, is the "interest" pressure from various groups. Trade association peer groups, who are not only affected and conforming but in many cases leading the way, are apt to take a dim view of any recalcitrant individual member who makes trouble for himself and other members as well. Unions, already dissatisfied with the Act in many ways and particularly critical of the relinquishment of federal administration of the national legislation to state enacted legislation and subsequent administration, are certain to watch the progress, or lack of it, closely. With the passage of time, as the Act becomes better known, public and citizen interest can also become critical.

Employers must maintain records of each job-related accident or illness. These records must be available for inspection purposes. A sampling from various selected industries goes to Washington, D.C. for study, analysis and evaluation. One annual form and summary must be posted for all employees to see. A system of fines and penalties has been devised as a deterrent measure for business that may tend to disregard the compliance specifications of the Act.

One far-reaching application and implication of the Act is that it involves the individual worker in several interesting aspects. The employee has both a right and a responsibility. He, or employee representative, can "request" an inspection, and do so with anonymous immunity. In the present Williams-Steiger Act, he cannot be cited or fined. Employees must be represented on all inspection tours and resultant conferences with management by a worker of their choice. Employees have "due recourse" or "due process" similar to employers. Employees can challenge the Secretary's decision if they feel his action is inimical or harmful to them.

Many employers become so involved in the employee rights aspect of the Act that they overlook and don't believe several positive factors in their favor. One of these is the responsibility of the employee. He must "comply with the safety and health standards applicable to the individual workplace." While an employer is charged and faced with the responsibility of enforcing a particular regulation, he does not have to keep an employee who is in constant violation of the employee responsibility. Such an employee can be discharged.

Also there is no collective bargaining issue here unless the Act, its Congressional intent, meaning, and specific terminology are completely subverted. Neither the employer, the employee, or the union, can abrogate their separate and collective responsibility under the terms of the Act. The employer's responsibility is plainly and simply stated, and the employee's responsibility is plainly and simply stated. The union, as the bargaining agent or representative of the employee in the middle between employer and employee, is faced with consideration of both sides of the coin.

As a matter of observable fact, the Act involves everybody who *is* involved about as totally and responsibly as can be set down on paper. Employers, employees and unions alike are bound together in this package in a fashion where no one member of this triumvirate can accuse the other without first tending to its own responsibility.

A union, for example, cannot argue the case of an employee proven to be in violation of a standard, and proven to have been given a fair and equitable opportunity to mend his ways, and failing that discharged by the employer. It cannot argue the case, that is, without defeating its own cause—the Act itself.

It is important, then, for the executive officer who may go no farther than these introductory pages to be sure that one individual, preferably a safety director, is responsible for the Act and to see that a safety committee is organized to help get the job done. The safety committee must understand what its functions and responsibilities are as they relate directly to the Act. It must be recognized that an already established safety committee, or safety director, may tend to follow already established lines of procedure and action. They must be made to see the fallacy of that immediately.

In all probability they will need to make a 180-degree turn. They should focus on the Williams-Steiger Occupational Safety and Health Act of 1970 and its corresponding state legislation. Practically speaking that is where all the industrial safety and health action is and will be.

Philosophically speaking, the Act is an excellent base for an effective and in many ways profitable and comprehensive safety and health program—a comparatively new experience for great numbers of business, industry and political subdivisions.

A last word of explanation. It is difficult, if not almost impossible, to write a definitive handbook or guideline on an Act that has two bases—federal and state. There are bound to be some discrepancies and differences. If, however, you and your personnel use the parent Williams-Steiger Occupational Safety and Health Act of 1970 as your informational source and specific guideline, you can be certain that your state Act will basically parallel OSHA—1970.

State Acts must be at least as effective. State administration of the program is monitored for three years by the Occupational Safety and Health Administration. If the state plan is ineffectual it can be removed and the federal plan reinstated. States, without exception, have developed their own plans with federal funds allocated to them for that purpose. The cost of administration of state programs is supported by the federal government on a 50-50 matching fund basis. All of it, plainly stated, is a kind of covert control mechanism difficult to ignore.

Although state Acts and plans may vary somewhat, if you substitute the state agency, director, compliance officer and due process for the federal terminology you will have a basic understanding of the Act.

CHAPTER THREE
Readthrough of the Act
— Salient Points

1. Employer "shall" provide a safe and healthful workplace for all employees.
2. Employer "shall" comply with safety and health standards.
3. Employee "shall" comply with safety and health standards.
4. "Standards" were issued 29 May, 1971, Federal Register, compiled from all federal contracts Acts, ANSI and NFPA standards—known as consensus standards. Changes, new standards, clarifications are constantly being made, issued and formulated. All are noted and brought up to date via the Federal Register, and via publication from OSHA when available.
5. Compliance inspections will be made by Federal compliance officers, by state inspectors deputized and trained as Federal Compliance officers, and, when state plan is in effect, by state compliance officers.
6. Citations will be issued on an "alleged" basis, if an employer or an employee is found in "violation" of any one of the standards applicable.
7. Employer can be assessed fines and penalties:
A. $1,000 for a serious violation.
B. $1,000 a day for each day, each violation, past the specified abatement period.
C. $10,000 for "failure to exercise reasonable diligence" in detect-

ing violations or willfully or repeatedly ignoring any of the obligations, standards, rules, orders or regulations of the Act.

D. $10,000, six months in prison, or both for willful violations of any standard, rule, regulation or order that causes death to any employee.

E. $20,000, one year in prison or both for subsequent willful violation that causes death.

F. $10,000 for false statement, representation or certification in any document, file or record maintained.

G. $1,000 or six months or both for advance notice of inspection without authorization (e.g., any warning ahead of time or notification of inspection by an unauthorized source).

H. $1,000 each violation of posting requirements.

I. $5,000-$10,000 and, or 3-10 years in prison for killing, assaulting or resisting certain Federal law enforcement officers in performance of officers in performance of official duties.

All fines are paid to the U. S. Secretary of Labor for deposit into U. S. Treasury, and accrue to the United States.

Note: The above are maximum. Fines can be, and frequently are minimal ($50.00, $100.00, etc.).

8. Reduction of fines (Fines can be reduced as much as 75% for nonserious violations, and 50% for both nonserious and serious violations.

Reductions are awarded on the basis of good faith, employer's attitude and evidence of willingness to improve safety performance, history of the employer, and size of the business.

9. Citations are issued for:

A. *Imminent Danger.* "Any conditions or practices in any place of employment which are such that a danger exists which could reasonably be expected to cause death or serious physical harm immediately or before the imminence of such danger can be eliminated through the enforcement procedures otherwise provided through this Act."

B. *Serious.* "If there is a substantial probability that death or physical harm could result from a condition which exists, or from one or more practice, means, methods, or processes which have been adopted, or are in use."

C. *Nonserious.* "A violation is not a serious one if an employer through the exercise of reasonable diligence, did not, or could not discover it."

D. *De Minimis.* "No direct or immediate relationship to safety or health." A notice that details the conditions and circumstances of the violation, and prescribes the means for correcting it.

Note: The citation, which forms the basis for subsequent enforcement proceedings, should describe "with particularity" the nature of the violation and should note the provision, standard, rule, regulation or order which has allegedly been violated. The citation should also fix a "reasonable" time period in which the employer must correct the violation.

10. Notice of penalty

Employer must be notified by *certified mail* within "reasonable time" after the termination of such inspection or investigation what the penalty is, if any, and that he has 15 working days after receipt of the notice within which to notify the Secretary if he wishes to contest the citation or proposed assessment of penalties.

The 15-day working period begins on the first working day after receipt of the notice. It does not include Saturdays, Sundays or Federal holidays.

If there is no contest within that 15-day period or no employee or employee rep files notice with the Secretary alleging that the time fixed in the citation for the abatement of violation is unreasonable, the citation and penalty are a final order and not subject to review.

11. Employer/employee/employee representative can appeal to Occupational Safety and Health Review Commission. If the employer wishes to contest a citation or proposed assessment of penalty, or if an employee or employee rep files notice alleging that the period of time for abatement of the violation is unreasonable, the Secretary must notify OSHRC, and the Commission must provide an opportunity for a hearing.

The Commission then issues an order, based on findings of fact, affirming, modifying or vacating the Secretary's citation or proposed penalty, or prescribing "other appropriate relief." The order becomes final 30 days after issue.

If an employer can demonstrate that he has tried, in good faith, to comply with the abatement requirements of the citation and that failure to comply is due to factors "beyond reasonable control" the Secretary after an opportunity for a hearing, shall issue an order affirming or modifying the abatement requirement in the citation. Affected employee representatives of affected employees shall be provided with an opportunity to participate as parties to the hearing.

12. Informal conference. Any employer or employee affected by a citation, notice of proposed penalty, or assessment of penalty, may request an informal conference with the Regional Administrator, or Solicitor, or the representative of either, to negotiate a re-inspection, or otherwise achieve a settlement of the issues. All parties may be represented by counsel. The conference does not serve as a stay of the 15-day notice of contest requirements.

13. The employer has further right of appeal to U. S. Court of Appeals. He can contest an order of the Occupational Safety and Health Review Commission by filing a petition with the U. S. Court of Appeals for the circuit in which the violation allegedly occurred, the circuit in which the employer has his principal office, or in the Court of Appeals for the District of Columbia.

14. Record keeping, Reporting. The record keeping began on 1 July, 1971, and will be prepared on a calendar year basis. Regulations require each employer covered by the Act to maintain three sets of records: (A) a log of all "recordable" (see below for definition) injuries and illnesses, (B) a supplementary record of illnesses and injuries, and (C) an annual summary.

A. Log of recordable injuries and illnesses (Form 100)

(1) Each employer shall maintain in each establishment a log of all recordable occupational injuries and illnesses for that establishment, except that under the circumstances described in paragraph 2 below, an employer may maintain the log of occupational injuries and illnesses at a place other than the establishment.

Each employer shall enter each recordable occupational injury and illness on the log as early as practicable but no later than 6 working days after receiving information that a recordable case has occurred.

For this purpose OSHA Form No. 100 or any private equivalent may be used. OSHA Form No. 100 or its equivalent shall be completed in the detail provided in the form and the instruction contained in OSHA Form No. 100. If an equivalent to OSHA Form No. 100 is used, such as a printout from data-processing equipment, the information shall be as readable and comprehensible to a person not familiar with the data-processing equipment as the OSHA Form No. 100 itself.

(2) Any employer may maintain the log of occupational injuries and illnesses at a place other than the establishment, or **by means** of data-processing equipment or both, under the following circumstances:

(a) There is available at the place where the log is maintained sufficient information to complete the log to a date within six working days after receiving information that a recordable case has occurred.

(b) A copy of the log, complete and current to a date within 45 calendar days, which reflects separately the illness and injury of that establishment, be available.

B. Supplementary Record (Form 101)

In addition to the log of occupational injuries and illnesses, each employer shall have available for inspection at each establishment within six working days, after receiving information that a recordable case has occurred, a supplementary record for each occupational injury or illness, for that establishment.

The record shall be completed in the detail prescribed in the instructions accompanying OSHA Form No. 101. Workmen's compensation, insurance, or other reports are acceptable alternative records if they contain the information required by OSHA Form No. 101. If no acceptable alternative record is maintained for other purposes, OSHA Form No. 101 shall be used, or the necessary information shall be otherwise maintained.

C. Annual summary (Form 102)

(1) Each employer shall compile an annual summary of occupational injuries and illnesses for each establishment. Each annual summary shall be based on the information contained in the log of occupational injuries and illnesses for the particular establishment. OSHA Form No. 102 shall be used for this purpose, and shall be completed in the form and detail as provided in the instructions contained therein.

(2) The summary shall be completed no later than one month after the close of each calendar year, beginning with the calendar year, 1971.

(3) Each employer, or the office or employee of the employer who supervises the preparation of the annual summary of occupational injuries and illnesses, shall certify that the annual summary is true and complete. The certification shall be accomplished by affixing the signature of the employer, or the officer or employee who supervises the preparation of the annual summary to the lower right hand corner of the annual summary, certifying that it is true and complete.

(4) Each employer shall post a copy of the establishment's summary in each establishment. The summary shall be posted no later than 1 February, and shall remain in place for 30 consecutive calendar days thereafter.

15. Recordable occupational injuries or illnesses: Those injuries or illnesses which result in fatalities, or lost workdays, and those which do not involve lost workdays but require transfer to another job, termination of employment, or medical treatment other than first aid are recordable. Cases involving loss of consciousness or restriction of work or motion, including any diagnosed occupational illnesses which are not reported to the employer, but are not classified as fatalities or lost workday cases are also included.

A. Medical treatment: treatment administered by a physician or registered professional personnel under standing orders of a physician.

B. First aid: one-time treatment and any follow-up visit for the purpose of observing minor scratches, cuts, bruises, splinters, and similar situations which do not ordinarily require medical care. Even when first aid is administered by a physician or registered professional personnel, it is not considered medical treatment.

C. Lost workdays: the total number of days, (not necessarily consecutive) following, but not including, the day of injury or illness that an employee was unable to perform all or any part of his normal assignment because of the job-related injury or illness.

D. Establishment: a single physical location where business is conducted, or where services or industrial operations are performed.

Records must be kept in each establishment for five years, following the end of the calendar year to which they apply. If ownership changes, the new owner is responsible for portion of year in which he assumed ownership. The previous records, however, must be kept for the time specified.

Records must be available for inspection and copying by Compliance Safety and Health Officers (CSHO), representatives of HEW, Bureau of Labor Statistics, and state compliance officers where state has been awarded jurisdiction over safety and health inspections or statistical compilations.

16. Reporting: Within 48 hours, employers must report to the nearest office of the Area Director:

A. occupational accidents or illnesses fatal to one or more employees.

B. occupational accidents or illnesses which result in the hospitalization of five or more employees.

The report may be oral or written but must detail the circumstances of the accident, the number of fatalities, and the extent of any injuries. The Area Director may then require any additional information he deems appropriate.

17. Falsification or failure to keep records or reports: **Whoever** knowingly makes any false statement, representation, or certification in any application, record, plan or other document filed or required to be maintained pursuant to this Act shall upon conviction be punished by a fine of not more than $10,000, or by imprisonment for not more than six months, or both.

Failure to maintain records or file reports required by this part, or in the details required by forms and instructions issued under this part, may result in the issuance of citations and assessment of penalties.

Any employer who wishes to maintain records in a manner different from that required by this part of the Act may submit a petition containing this information to the Regional Director of the Bureau of Labor Statistics wherein the establishment involved is located. The petition shall include:

(1) The name and address of the applicant.

(2) The address of the place or places of employment involved.

(3) Specification of the reasons relief is sought.

(4) A description of the different record keeping procedures proposed by the applicant.

(5) A statement that the applicant has informed his employees of the petition by giving a copy thereof to their authorized representative and by posting a statement giving a summary of the petition and by other appropriate means. A statement made pursuant to this shall be posted in each establishment. The applicant shall also state that he has informed his employees of their rights.

(6) In the event an employer has more than one establishment he shall submit a list of the states in which such establishments are located and the number of establishments in each such state. In the further event that certain of the employer's establishments would not be affected by the petition, the employer shall identify every establishment which would be affected by the petition and give the state in which they are located.

(7) Any employer with eight or less employees is only required to report (a) an occupational fatality, or (b) an injury resulting in multiple hospitalization.

18. Citation posting requirements: These must be prominently posted by the employer at or near the place where the violation occurred—a place where employees will see the notice. They must remain for three days or until the hazard is abated, whichever is later, or until the Commission issues a final order that the citation should be dismissed.

The employer may also post a notice in the same location stating that the citation is being contested before the Commission.

19. Employee rights and responsibilities

A. An employee or employee representative can "request" an inspection. This can be initiated by a telephone call to the Area Director, but ultimately it must be submitted in writing.

The request must state that the employee or employee representative believes that a "violation of safety or health standard exists that threatens physical harm or that imminent danger exists" in a place where work is performed. The notice must set forth with "reasonable particularity" the conditions for which the inspection is being requested. It must state also whether it concerns an imminent danger and what steps the employee or employee representative has taken to rectify the situation.

Although the notice to OSHA must be signed by either the employee or employee representative, the name or names of employees referred to therein will not appear in the copy of the notice or any report of information published pursuant thereto. The Area Director or Compliance Officer must provide the employer with a copy of the notice no later than the time of the inspection.

If the Secretary determines, as a result of the employee allegation there are reasonable grounds for an inspection, he shall initiate a special inspection as soon as practicable to determine whether such violation or danger exists. The inspection need not be limited to matters referred to in the request for an inspection. If the Secretary determines that grounds for such an investigation do not exist, he must so indicate to the employees or their representative.

In the course of such an inspection, or any inspection, employees can call the attention of the compliance officer to violations of any safety and health standards.

B. Employee representatives can accompany compliance officers on inspection tours of workplaces.

C. Employee representatives are to be included in the Compliance Officer's report to and conference with the employer after the inspection tour.

D. Employees may communicate with the Secretary in connection with any proceeding under the Act.

E. Employees may participate as parties to any hearing, contest any action of the Commission, or contest the period of time allowed for the abatement of a violation.

F. Employers must keep employees informed of their protections and obligations under the Act including the provisions of applicable standards.

G. The Act provides safeguards for employees who feel they have been discharged from their jobs or discriminated against for having participated in any proceeding under the Act.

H. Employees may serve on the National Advisory Committee on Occupational Safety and Health, in order to advise and consult with the Secretary on all aspects of safety and health programs.

I. Employees must be represented on advisory standard-setting committees.

(1) Each standard promulgated under the Act must assure to the extent feasible that no employee will suffer material impairment of health or functional capacity even if such employee has regular exposure to the hazard dealt with by such standard, for the period of his working life.

(2) Such standards must include requirements for (a) labels or other appropriate forms of warning to alert employees to all hazards to which they are exposed, (b) suitable protective equipment, (c) monitoring of exposure to toxic materials, and (d) medical examinations and tests, including employee access to such records.

J. "Any person adversely affected or aggrieved" by an order of the Commission, can seek "judicial review" by filing a petition with the U. S. Court of Appeals for the circuit in which the employer has his principal office, the circuit in which the violation allegedly occurred, or in the Court of Appeals of the District of Columbia. (This means employee or employee representative.)

K. Should the Secretary of Labor "arbitrarily or capriciously fail to seek relief in an 'imminent danger' situation, any employee who feels he may be injured as a result of such failure, or any authorized employee representative may seek a writ of mandamus in the appropriate U. S. District Court, requiring that the Secretary initiate proceedings for a temporary restraining order and any other relief which can be appropriate."

L. No employee shall be subjected to medical examination, immunization or treatment contrary to his religious beliefs, except where such examination or treatment is necessary to protect the health and safety of others.

M. Employees must have access to any copies of the Act and/or applicable safety and health standards furnished to employers by OSHA. If an employee is denied permission to review the Act or standards during working hours, he must be afforded access to the documents after work on the same day the request is made or at another suitable time.

N. Employees cannot be fined or cited.

O. "Each employee shall comply with the occupational safety and health standards and all rules, regulations, and orders issued pursuant to this Act which are applicable to his own actions and conduct."

20. Employer rights and responsibilities

A. "Each employer shall furnish to each of his employees employment and a place of employment which are free from recognized hazards that are causing or likely to cause death or serious physical harm to his employees, and shall comply with the occupational safety and health standards promulgated under this Act."

B. The employer can enforce the compliance responsibilities of his employees.

(1) He must determine his own enforcement program. Nothing specifies what it should be or how it should be done, except to indicate that the responsibility is his. Such an enforcement program *could* include: (a) a warning letter to employee on first violation; (b) a three-day layoff on second violation; and, (c) discharge on third violation. A careful reading of the Act reveals that the backup for this is the Act itself, not the Compliance Officer, not the Area or Regional Director, nor the Secretary himself. This applies equally to management and labor, employer and employee —and is not a means for jurisdictional argument or contract negotiation, or collective bargaining.

C. The employer shall conduct, proffer or make possible those educational programs necessary for the achievement of some specific requirements under the Act, (see noise abatement and hearing conservation programs heading) and also the goals and objectives of the Act generally.

D. The employer can appeal citations, assessments, penalties, and decisions to OSHRC, U. S. District Court of Appeals, and, if so desired, to the Supreme Court.

E. The employer can apply for and secure valid and justifiable variances.

F. The employer can, on his own and through industry associations, seek and suggest valid and justifiable changes in the standards.

G. The employer can make application to the Small Business Administration for loans to assist him in the purchase of gear, equipment, engineering design, building, etc. related to the achievement of the goals and objectives of the Act and its compliance requirement.

H. The employer can, by attitude, good faith, willingness to improve safety performance, and past history, secure reductions in assessments, fines and penalties.

21. Variances: The Secretary may grant variances, temporary or extended, from occupational health and safety standards (or provisions of standards) in situations where an employer:

A. must increase his staff or enlarge his facilities in order to comply with the standard,

B. can provide working conditions as safe and as healthful as those he would be required to maintain if he complied with the standard,

C. is taking part in a worker safety or health experiment.

When an employer is unable to comply with a standard by its effective date, he may petition the Secretary of Labor for a temporary order granting a variance from the standard. This is only granted after employees have been notified of the employer's request and are afforded an opportunity for a hearing. The Secretary, however, may issue one interim order which will be effective until a decision is made on the basis of a hearing.

The temporary order remains in effect until such time as the employer can come into compliance with the standard, or for one year, whichever is shorter, and may be renewed twice, as long as the employer continues to meet the requirements of the temporary order, *and* if the renewal application is filed at least 90 days before the expiration date of the order. No interim renewal of an order may remain in effect for longer than 180 days.

All temporary orders must prescribe the methods and practices the employer must adopt while the order is in effect, and they must set forth, in detail, his plan for coming into compliance with the standard. To qualify for a temporary variance, -an employer must establish that:

A. he is unable to comply with the standard due to unavailability of qualified personnel, required material or equipment, or because "necessary construction or alteration of facilities cannot be completed by the effective date,"

B. he is taking all possible steps to safeguard his employees against hazards covered by the standard,

C. he has an effective plan for coming into compliance with the standard "as quickly as practicable."

He must submit to the Secretary an application containing the following information:

A. specification of the standard or portion of the standard from which he seeks a variance,

B. a representation, supported by qualified persons, having firsthand knowledge of the facts represented, that he is unable to comply with the standard (or portion thereof) including the reasons for inability to comply,

C. a detailed statement of the measures he has taken and will take to protect employees against the hazard covered by the standard, together with the dates or target dates for such actions,

D. a detailed statement of the measures he has taken or will take to comply with the standard, together with the dates corresponding to such action and the target date for compliance with the standard,

E. certification that he has informed his employees of this application for variance by furnishing a copy of that application to their authorized representative, by posting a summary of that application, by specifying where employees might examine a copy, and by other appropriate means. (The certification should include a description of what the employer has done to inform his employees.)

22. Inspection procedures: Inspections must occur during regular working hours and at other reasonable times, within reasonable limits, in a reasonable manner, and may include private questioning of any employer, owner, operator, agent, or employee. In other words, "during regular working hours and in a manner which does not necessarily disrupt normal business procedures."

The Compliance Officer must, upon entering the workplace:

A. present credentials to the owner, operator, or agent in charge. If, however, within a reasonable time period, an inspector is unable to locate an "agent in charge," he may gain entry by presenting his credentials to any other employee, thus eliminating any time lapse during which an employer could cover up violations of the Act.

B. explain the nature, purpose, and the scope of the inspection, and

C. indicate the records he wants to review and the employees he wants to question.

Advance notice of any inspection not authorized by the Secretary is prohibited and is subject to a fine of not more than $1,000, or imprisonment of not more than six months, or both. Advance notice of the inspection, however, can be given and authorized by the Area Director or Compliance Officer in cases where:

A. such notice will enable an employer to abate dangerous conditions as quickly as possible,
B. the inspection will be conducted after regular business hours, or in circumstances where special preparations are necessary for an inspection.
C. notice is necessary to assure the presence of an employer and employee representative or personnel needed to aid the inspection,
D. other circumstances in which the Area Director or the Compliance Officer, after consultation with the Area Director, determines that notice would permit a more effective inspection.

With rare exception, no notice of inspection may be given more than 24 hours before the scheduled inspection.

An employer representative and an employee representative may accompany the Compliance Officer on his inspection. If there is no duly recognized employee representative, the inspector must consult with a reasonable number of employees regarding conditions of health and safety in the establishment. (He cannot appoint an employee representative as such.)

The right to accompany the Compliance Officer will be denied to anyone who interferes with a fair and orderly investigation. In certain instances the right to accompany the officer may be denied where security matters or trade secrets are involved. In workplaces where groups of employees are represented by different employee representatives, more than one employee representative may accompany the Compliance Officer, or he may be accompanied by different representatives during different phases of the inspection.

Upon completion of inspection, Compliance Officer must confer with the employer or his representative with the employee representative present to advise the employer of safety and health violations which have become apparent as a result of the inspection. The Area Director, after conferral with the inspecting Compliance Officer, sets the proposed amount of the fines, if any.

23. State participation: In keeping with the expressed purpose of assisting and encouraging the states "in their efforts to assume the fullest responsibility," the Act provides options for retention of state jurisdiction in the development of enforcement of health and safety standards. It provides further that even in areas where Federal standards have been established a state may submit its own plan for approval by the Secretary of Labor provided that such plan is at least as effective as the Federal program, and, meets certain specified criteria (see below).

The Act also provides for:

A. procedures permitting interim enforcement of state standards pending final approval of state plans,

B. Federal grants to facilitate the establishment of state programs in furtherance of Congressional safety and health goals, and

C. cooperation between Federal and state agencies in carrying out the objectives of the Act.

A state may have jurisdiction, under state law, over any occupational safety or health issue with respect to which no Federal standard has been promulgated. However, even when Federal occupational health and safety standards have been adopted, any state wishing to assume responsibility for development and enforcement of its own standards may draft and submit for approval by the Secretary of Labor a plan outlining the details of such a program (18h agreement, 18b plan).

The state's plan may deal comprehensively with the problems of health and safety, or may restrict itself to a single hazard or industry. However, any hazards or industries not covered by the state plan will continue to come under Federal jurisdiction.

Eight Requirements for Acceptance of State Plan

(1) Identify the state agency or agencies responsible for administering the plan throughout the state.

(2) Provide for the development and enforcement of health and safety standards which will be at least as effective as Federal standards promulgated under Section 6 of the Act.

(3) Provide for a right of entry and inspection of all workplaces subject to the Act, which is at least as effective as that provided in Section 8, and which includes a prohibition against advance notice of inspections.

(4) Contain assurances that the designated state agency or agencies will have the legal authority and qualified personnel for the necessary enforcement of such standards.

(5) Provide assurances that the state will devote adequate funds to the administration and enforcement of the standards.

(6) Contain assurances that the state, "to the extent permitted by its law," will establish and maintain an occupational safety and health program for all public employees of the state and its political subdivisions, which will be as effective as the program approved in the plan.

(7) Require employers to report to the Secretary in the same manner and to the same extent as if the plan were not in effect.

(8) Provide that the state agency will report to the Secretary in such form as the Secretary shall from time to time require.

Option to Enforce Federal Standards

After approval of the state plan, the Secretary of Labor may, but shall not be required to, exercise his authority to enforce Federal standards. He first determines whether the state plan is operating according to criteria. He does not make this determination until more than three years after state plan is operational.

If the plan is consistent with the criteria, the Secretary of Labor may relinquish his option to enforce Federal standards, and the state will perform those functions related to

(1) employer compliance with standards promulgated under the Act,
(2) inspections, investigations and record keeping,
(3) citations,
(4) procedure for enforcement,
(5) procedures to counteract imminent dangers, and
(6) penalties.

State Plans to be Evaluated

The Secretary is required to evaluate continually the manner in which a state is administering its plan. This evaluation is based on reports submitted by the state agency and information obtained through his own inspections. If the Secretary finds that any state, after due notice and opportunity for a hearing, has failed to comply substantially with any provision of its plan, he must notify the state agency that he is withdrawing approval of the plan.

Upon receipt of such notice by the state agency, the plan will cease to be in effect, and the Secretary may reassert his full authority under the Act.

The State can petition the U. S. Court of Appeals (within 30 days) for review of the Secretary's decision. Judgment of that Court can be subject to review by the U. S. Supreme Court.

Contents of 18h Agreement (State's application)

1. A brief description of each standard which will be covered by the agreement, including pertinent legislative or regulatory citations, a description of the safety and health issue covered by each standard, and a listing of the employments, hazards and establishments which will be covered.

2. The names, addresses and telephone numbers of the state agency or agencies responsible for administering and enforcing the state standard(s), in addition to the names, addresses and tele-

phone numbers of the heads of such agencies. (These requirements do not apply to any agency having only judicial or quasi-judicial responsibilities.)

3. A description of the enforcement program and procedures of each agency administering the standards, containing any specific information which the Secretary deems necessary.

4. The dollars and "approximate man-years" allocated within each responsible agency during the current and previous fiscal periods, including a description of the fiscal period.

5. Notice of the state's intention to submit a plan under Sec. 18b including the state agency's agreement to, (a) submit a notice of intention to file a Sec. 18b plan within 60 days following publication in the Federal Register of procedures for submission and approval of such plans, and (b) submit a plan within the prescribed 60-day period, which may be expected to meet the requirements for state plans with little or no modification thereof.

6. A brief description of the Federal standards dealing with the same issues as the standards which are the subject of the agreement.

7. Provisions under which the Secretary shall inform the state agency of his enforcement activities within the state with respect to the Federal standards referred to in the agreement and provisions requiring the state agency to make such report to the Secretary he may deem necessary with regard to activities performed under terms of the agreement.

8. Any other provisions the Secretary may deem "appropriate" under the circumstances of the standards issue involved.

CHAPTER FOUR
Safety Coordinator/Director

SELECTION

Until the passage of the Act the position of safety director was a neglected one. As a result truly qualified and competent safety personnel are now critically limited in number.

To overcome this, training programs, safety and health curriculi at the university level, enabled by grants from HEW-NIOSH are currently being developed. Numerous organizations conduct and sponsor specialized short courses in safety and health. It is going to take a long time, however, for the newly-trained supply to meet a very critical demand. The Act and its requirements will create literally thousands of new jobs across the country. For example, our forecast for the Region X area, Washington, Oregon, Idaho and Alaska, is 3,500 administrative and support jobs in the five years following 1973. If you multiply that by ten regions this could add up to 30—40,000 new jobs in safety and health.

The Act is said to embrace and cover some 62 million workers. This figure is the total within the federal administration of the Act. If we add to that workers who will be covered in addition in all political subdivision sectors—the federal by presidential decree, state, county and city by state administration of state legislated Acts—we might settle on some 120 million workers in total.

If we approximate the numbers of personnel engaged in the responsibility of safety, present and future, at 50,000, each safety and health administrator and support worker would be responsible for some 2,000 workers. If any qualified and competent safety professional were asked what he thinks can really be accomplished and achieved with 2,000 workers, his answer in all candor would be, "Not much."

There is no question that any company applying itself assiduously and intelligently along the lines suggested herein can *adequately* manage its safety and health problem within its own presently established structure. There is no question, equally, that adequate is not going to be enough.

To practically, economically and profitably achieve the requirements and specifications of the Act, any company of 100 employees or over is going to need to either (1) employ a full time safety director as such, or (2) use professional safety expertise on some kind of consulting basis. For the present, business and industry will have to make do with what it has and what is available.

The safety and health responsibility should be that of one person who is educated and trained for the job as much as is possible. Most of today's qualified safety directors, managers or coordinators, have learned by doing. Great numbers of them have simply been pressed into service when the need arose as the only company employee available at that time, sometimes with and sometimes without qualification. Very few of them have graduated from a college or a university with a major in safety and health as such. Some have college degrees in engineering, industrial hygiene, or nursing—the closest approximation to the specific safety requirement.

The really capable safety professional, the learn-it-the-hard-way, do-it-yourselfer, is already at work for a company which discovered long before the Act came into existence that loss control, accident prevention and safety were an integral part of profitable operation.

In selecting a safety coordinator, it would be best to select a candidate—perhaps from your safety committee—with a demonstrated ability to develop, organize, present and conduct a project or a program, somebody with a recognizable ability to communicate effectively, to get along well with his fellows, to secure their respect, and to handle the leadership role. No one individual will combine all the talents you are seeking, but someone will have capabilities in one or more of these areas. That individual should be the safety coordinator, and should be assigned the OSHA-1970 responsibility.

The newly appointed safety coordinator should be trained with whatever training and education programs are available. Classes in management and communication techniques are essential. Any courses in safety and health, on a continuing or extended education basis, with or without credit, are recommended. A curriculum at any kind of university or college level that is part of either an undergraduate or post graduate degree program would be an extremely desirable part of any educational or training program for any safety personnel.

The safety coordinator should be educated and trained as comprehensively and competently as the rest of the supervisory and management employees. He should also be placed on a report-direct basis to top-level management. Place him at this trainee level in some kind of line of management function and participation where the importance of the job is recognized and not second-rated.

If a safety and health director is hired who is already qualified by experience, background, training and education, the above suggestions still apply, but there are obviously some factors already in your favor. For example, he already knows his job, he is probably thoroughly acquainted with the Act, but this must be verified. He knows the business and the people in it. He usually knows what to do, how to do it, and if not, where to go and who to get to do it. He is usually well versed in the training, education and information roles involved in the safety business. He has probably attended most of the good courses, seminars, conferences and meetings, not only on safety but in management techniques and other disciplines. Another interesting fact about him is that he usually has a rather uniquely broad, across-the-board contact with a large segment of the business, industrial, governmental (national, state and local), medical, legal, insurance, professional safety, and community, life. However, he needs several things he has seldom secured prior to the passage of the Act. He needs to report directly to top management—president, executive vice president, owner. And he needs to be a part of top line management, of equal importance, status and income with that level of management. He needs to be heard on all issues related to safety and health programming and the successful management of it. No annual budgets should be established without considering what expenditures are needed to meet standards requirements of the Act. He should be required to give his opinions on those areas of function, operation and production directly related to his profession, expertise and responsibility.

Perhaps for the first time in your business you are going to need a competent, capable, well trained and educated, experienced, practical and toughminded, human-related and profit-oriented safety and health professional.

THE SAFETY DIRECTOR

The safety director, as capable and experienced as he may be, must realize that safety management has changed. Whether corporate director of safety for the largest company in the United States, director of safety for the biggest labor union, chairman of the federal safety council, or manager of the various safety councils, trade associations, industrial or business organizations around the nation, each person must see that what *has* been done is not enough. There is still a great deal to be accomplished, and safety directors should consider the following as they work.

1. Assess yourself and the problem you face as the leader, guider and director of the problem, challenge and opportunity the Williams-Steiger Occupational Safety and Health Act of 1970, or its state version, presents.
2. Do not be content with the old established ways. Use only those that apply, and develop new approaches, methods and means.
3. Become thoroughly, painstakingly acquainted with the Act. Look carefully at some of its implications, notably employee relations.
4. Make it work *for* you, your company, and your charges. Don't work against it, wasting time in wheelspinning complaint or criticism.
5. Be realistic about your ability. As remarkably well-informed as you may be, don't think you have all the answers.
6. Organize a safety and heath committee. Assess or reassess it in terms of (a) better representation from management, (b) broader representation from total company structure, (c) representation from unions involved, (d) better representation from and contact with, employees, and (e) reorientation of goals and objectives.
7. Develop your own plans, programs, ways and means, costs involved, and annual budget. Submit this to management as a matter of record and recommendation.
8. Maintain and keep a cost-accounting record of your department and all costs related to safety and health programming.
9. See that the OSHA-1970 and state version story is told to all company personnel. Do this by:
A. flip chart presentation and other visuals,
B. posting of notices, memos, records and reports,
C. publication of information relevant to the Act,
D. meetings with employees, and

E. securing guest experts, authorities, and officials to speak or make presentations on the Act.

10. Attend all relevant meetings, conferences, congresses, and seminars on the Act.

11. See that key members of safety and health committees and representatives from management attend as many of these as practicable.

12. Conduct, or cause to be conducted, specialized courses on safety and health, management and communicating techniques, motivation, instruction and training on the Act and any of its specific phases.

13. Take whatever accredited safety and health courses you can.

14. Run a very careful, specific and detailed assessment of the standards applicable to your company. Organize and direct "self-inspections." Determine what needs to be done, and get it done.

15. Determine and list all available information sources relevant to the Act.

16. Subscribe to any informational services you consider useful.

17. Compile a list of OSHA and state personnel, names, addresses, and telephone numbers. Know who they are, what they do, and how they can be of assistance.

18. Get to know these people by other than just name, rank and serial number. They have a job to do, but they are also human beings.

19. Join and participate in professional societies, **safety councils,** trade, business and industrial associations. Learn about your fellow professionals, their projects, problems, ideas, plans, and programs.

20. Participate in panels and discussions, make speeches and presentations, and write papers and articles on safety and health.

21. Insist on a report-direct line from you to the principal executive officer in your company.

CHAPTER FIVE
Safety and Health Committee

The Safety and Health Committee should consider the following:
1. Secure as a member the top executive officer in your company. If this is not possible, obtain someone as near the top as you are able.
2. Be sure there is a report-direct line to the top executive.
3. Secure as broad a representation from staff (administrative) and line (operations) management and personnel, as you can obtain (Vice presidents, Directors of Finance, Sales and Marketing, Advertising and Public Relations, Purchasing, Plans and Programming, Engineering, Personnel, Industrial Relations—Managers, Superintendents, Supervisors, Foremen, Leadmen and Employees.)
4. Establish goals and objectives, ways and means of achieving them, meeting dates, and target dates where feasible.
5. Publish this list along with other pertinent data for total company personnel distribution and dissemination.
6. The primary goals and objectives should include:
A. planning, developing, organizing and conducting a safety and health program centered on and directed towards OSHA-1970.
B. include in that plan the rest of the safety and health spectrum —off-the-job, traffic, home, recreational—within practical achievable goals.
7. Other goals:
A. assessment of standards, both federal and state. What are they? How do they apply? What is now being done? What needs to be done?

(1) unstaple the Federal Register, tab it and organize it by sections.
(2) assign these sections to safety and health committee members for evaluation and recommendation on necessary steps to be taken.
(3) determine what needs to be done to comply with the Act.
(4) take whatever action, within company policy and safety and health committee responsibility, is necessary.

B. Become thoroughly acquainted with the Act, with its basic specifications and requirements.

C. Develop and help maintain the Safety Log. (This can be a responsibility of the Committee but is probably the responsibility of the Safety and Health Director/Coordinator/Consultant, if you have one.)

D. Develop, define and delineate duties of all members of the Committee.

E. Publish, post and distribute all relevant information and data developed by the Committee.

F. Attend all meetings, conferences, congresses, seminars on OSHA-1970. See that as many members of the Committee attend as practicable and feasible.

G. Conduct self-inspections regularly. Report all findings to the Committee. Make corrections necessary if possible, and/or report corrections necessary to management.

H. Help develop the enforcement measures necessary to aid employees in effectively managing their responsibility of the "shall comply" requirement of the Act.

I. Help enforce those measures by suggestion, example and leadership, and by reporting.

J. Help develop, conduct and secure programs of information dissemination, training and education for the entire company structure, management and employees.

K. Take specialized courses in safety and health education, instruction and training.

L. Join professional health and safety societies.

M. Join and participate in community, industry, association health and safety councils/committees.

N. Participate as a speaker or panel member at safety meetings other than your own.

Note: The Safety and Health Committee is *not* responsible for compliance under the Act. The employer has the principal responsibility in two phases: "Each employer shall furnish to each of his employees, employment, and a place of employment, which are free from recognized hazards that are causing, or likely to cause, death or serious physical harm to his employees," and he "Shall comply with the occupational safety and health standards promulgated under the Act."

The employee has one principal responsibility: he "Shall comply with safety and health standards as they relate to the workplace."

The Safety and Health Committee, however, can and should be a prime force in helping achieve those compliance requirements for both employer and employee.

Even though a company has a safety and health director, it should without exception have a safety and health committee. If this is not the case, the company has:

1. not grasped the full significance of the Act, its meaning, application and implication,
2. made the task of achieving compliance much more difficult, if not impossible,
3. failed to make use of "many hands" available and necessary to get the job done,
4. missed an excellent employee relations potential, and
5. aborted the built-in opportunity to achieve the goals and objectives of an effective safety and health program as practically, economically, and profitably as possible.

No Safety and Health Director can accomplish this task without a functioning Safety and Health Committee.

CHAPTER SIX
Safety Log

If a safety log is already kept, it should be examined within the context suggested here. This is a first step listed in this guideline for the very important reason that it is a recorded keystone for an effective safety and health program. The items listed below are some suggestions for inclusion in the safety log.

1. Letter of intent (see sample on Page 37). This is a simple letter of intent about the Act, signed by your executive officer, prominently posted, and mailed to all employees.
2. Sample letters of results of audiometric testing of employees (see Noise Abatement and Hearing Conservation Program section.)
3. Names, responsibilities and job outlines of safety and health directors/coordinators/consultants.
4. Names, responsibilities, departments represented, of Safety and Health Committee.
5. Published notice of Safety and Health Committee's objectives and goals, methods and means, regular meeting dates and whatever other information is pertinent.
6. This should originate from a formal resolution by the Safety and Health Committee, which spells out the above specific facts.
7. Copies of all record keeping and reporting pertaining to the Act, or any safety and health programming/function/activity.
8. Records of attendance at safety and health meetings of any description, copies of programs, if possible, and copies of any briefs/summaries prepared as a result.

9. Results of any compliance inspection with dates and details. Copies of any memos, letters, dissemination of information of any kind.
10. Results of self-inspection.
11. Record of any action taken:
A. engineering changes
B. improvement or check-out on specific situations
C. variances applied for or secured
D. appeals to Board of Review or others
12. Enforcement measures adopted: what they are, why they were taken, and what the results were.
13. Education and training measures.
A. classes/courses instituted and conducted in-plant
B. classes/courses attended out-plant
14. Institution of noise abatement and hearing conservation program.
A. when started, deadlines
B. details on areas and workers involved
C. audiometric testing summary
D. engineering approach
E. administrative controls
F. employee personal protective devices
15. Purchase of any equipment, gear, devices, mechanism, or programs.
A. cost
B. what purpose
C. what accomplished
D. for whom issued
16. Membership in professional societies, associations.
A. offices held
B. projects participation
C. special recognition and awards
17. Company participation in, sponsorship of, meetings/conferences/congresses/seminars.
18. Company participation in industry-wide, association, society, safety and health projects, research, scholarships.
19. Published articles or papers by company personnel.
20. Speeches, participation on panels, assistance in organization and conduct of safety and health meetings/congresses/conferences/seminars.
21. Inclusion of all letters, memos, notices of action taken, action recommended, notification, and notices.

LETTER OF INTENT

Dear fellow employee:

The Williams-Steiger Occupational Safety and Health Act of 1970 is now in effect and has been since August 27, 1971. It applies directly to all of us—to us as an employer and to you as an employee. The Act is a "voluntary compliance" law, seeking cooperation from management and labor, employer and employee. Our compliance with the Act specifies that we "shall" provide a safe and healthful work place for every employee. This is accomplished by meeting a set of safety and health standards specified in the Act. Your compliance with the Act requires that you "shall" meet those standards by individually employing safe and healthful work habits. This is also specified in the Act.

Our position and intent on this new and comprehensive Act is to cooperate to the fullest of our ability and to meet its voluntary compliance specifications as successfully and capably as we can possibly manage. You will be hearing more about this in detail as our program for voluntary compliance is devised and implemented. I know I can count on your support, cooperation and assistance in helping our company achieve the much-to-be desired goals and objectives of the Williams-Steiger Occupational Safety and Health Act of 1970.

Thank you for listening.

CHAPTER SEVEN
Noise Abatement and Hearing Conservation Program

OCCUPATIONAL NOISE STANDARDS
(Reference: Bulletin 334, U. S. Department of Labor)

These standards basically require that a noise abatement and hearing conservation program be instituted by an employer in any workplace where noise levels exceed 90dBA (90 decibels on the "A" scale). The program specifies that:

1. engineering design/controls/measures be taken by the employer to achieve sound levels throughout the work place that do not exceed 90dBA.
2. administrative means be employed in the placement of employees in work areas, which in combination of several exposures above and below 90dBA, add up to a specified number of working hours, or portions thereof, within exposure times and levels, averaging out to 90dBA for an eight-hour working day.
3. personal protective devices (ear plugs, muffs) be made available, used and required for all employees working or entering, either steadily or infrequently, those areas exceeding 90dBA.
4. all employees who either steadily or infrequently work or enter areas exceeding 90dBA be given regular audiometric tests to determine threshold hearing, and that those records be available for inspection by any compliance officer representing either OSHA, or the state agency administering the Act.

5. periodic noise surveys be made in those areas where levels exceed 90dBA to determine what they are in the first place and to check the results of any engineering changes or any change in the noise level.

6. the objective, and end result, as specified in the standards, is that any noise level exceeding 90dBA must be engineered out within some kind of specified time period, insofar as this is feasible. "Feasible" as defined by the guideline is "capable of being done, accomplished or carried out; capable of being dealt with successfully."

7. Noise Surveys

A. A noise survey of each area in the plant in which sound levels exceed 90dBA shall be made at least once a year to ensure that sound levels have not increased above those originally existing.

B. The survey may also establish that noise levels in some area have been reduced to levels below 90dBA and thereby justify discontinuing application of requirement for administrative controls, ear protection, and audiometric tests of individuals in such areas.

C. Noise survey of an area is recommended whenever a change is made in either equipment or type of operations so that significant changes in noise level will be acted upon immediately.

D. Tests of noise levels will be made with a sound level meter on the A-scale, slow response.

E. The sound level meter used will meet ANSI standards.

F. Record of noise survey will show: instrument used, date, time and location of such tests, machinery or equipment generating the noise, name of person making the test.

G. Test records of such surveys will be kept readily available for inspection for one year or until a subsequent survey is made, if done more frequently.

H. The noise survey will be done by an insurance carrier, consultant, representative of a state labor or health department, or by a qualified individual designated by the company.

The following measures, though required and essential, are in a sense substitutes.

1. Administrative Controls

If noise cannot be reduced to permissible intensities by engineering controls, administrative controls should be developed in order to limit duration of workers' exposure to noise levels above 90dBA to the limits explained in Bulletin 334. Some examples:

A. Arrange work schedules so that employees working the major portion of a day at or very near to the 90dBA limits are not exposed to higher noise levels.

B. Ensure that employees who have reached the upper limit of duration for a high noise level work the remainder of the day in an environment with a noise level well below 90dBA.

C. Where the man-hours required for a job exceed the permissible time for one man in one day for the existing sound level, divide the time/work among two, three or as many men as needed, either successively or together, to keep individual noise exposure within permissible time limits.

D. If less than full-time production of a noisy machine is needed, arrange to run it a portion of each day, rather than all day for part of the week.

E. Perform occasional high level noise producing operations at night or at other times when a minimum number of employees will be exposed.

2. Personal Protective Equipment

"If such control fails to reduce sound levels within the levels of the specification, personal protective equipment shall be provided and used to reduce sound levels within the levels permissible."

A. This is considered an interim measure. There are few cases where the use of such equipment is acceptable as a permanent solution.

B. Regulations require both the *provision* and *use* of personal protective equipment. How this is accomplished is up to the employer.

(1) Educational and promotional programs are to precede the initiation of required use of such equipment, and "continue until 100% acceptance by employees."

(2) The absence of an observable high proportion of use constitutes a "violation of the regulation."

AUDIOMETRY

Audiometric tests are to be made of all individuals working steadily or infrequently in areas in which the noise level is 90dBA or above. Audiometric tests may be made as frequently as prescribed by the plant regular or consulting physician, but only under very special conditions should they be made less than once a year. They can be made in a doctor's office or elsewhere outside the plant, providing test facilities, techniques and records are equal to, or exceed, the minimum requirements described below:

A. Test booth or room must meet current ANSI standards.

B. Tests shall consist of an air conduction octave band analysis, ANSI standards, 500, 1,000, 2,000, 3,000 and 4,000, minimally.

C. Audiometric tests must be conducted by persons trained and skilled in audiometric testing.

D. Audiometers must meet the specifications for limited range, pure tone audiometers (ANSI standards).
E. Audiometers must have a certificate of calibration, issued by someone qualified to so certify, before being placed in use.
F. Audiometers shall be calibrated each year thereafter.
G. The current certificate attesting to such calibration shall be ready for inspection.
H. Audiometers shall be subjected to a biological check, preferably once a week but at least once a month, or before each use of the instrument if it is used less than once a month. A log of these checks shall be maintained and be available for inspection.

RECORDS

A. A record of each audiogram made on each individual shall be available for inspection.
B. Records of audiometric tests shall indicate which standard they are based on.
C. Complete records on each employee required to be tested shall be retained for one year following termination of employment or transfer to an area in which noise levels above 90dBA do not exist.
D. Records will be examined for evidence of any deterioration of hearing acuity and of action taken to prevent further deterioration in those employees found to suffer some loss of acuity.
E. Conclusions as to effectiveness of control measures taken will be based on examination of a significant number of audiograms and not upon the basis of one or two cases.

SELECTION OF PERSONAL PROTECTIVE EQUIPMENT

Cotton stuffed in the ears is not acceptable, but fine glass wool is. Wax-impregnated cotton, properly inserted in the ear, if fresh daily and properly used is acceptable. Properly fitted ear-plugs, fitted to the individual, issued by a trained person under the direction of a physician or the physician himself, and frequently checked are acceptable.

Also acceptable are ear muffs. These can be issued by any designated person in the plant. Long hair and spectacle or goggle temples reduce attenuation. The ear protector used must achieve the reduction of the noise level to the permissible specification.

HEARING CONSERVATION PROGRAM

"In all cases where the sound levels exceed the values specified, a continuing, effective hearing conservation program shall be administered." "Continuing" means that the program will be in effect and in use as long as noise levels above 90dBA occur in the plant. "Effective" means that exposed employees will not suffer continuing deterioration of hearing acuity because of their exposure, but that incipient loss of hearing will be detected and necessary steps taken to prevent any further deterioration before serious hearing loss has occurred.

A. Where the sound level in a working area has not been reduced to 90dBA or below by engineering means, and reliance must be placed on administrative controls to limit duration of exposure, or on ear protection to reduce the sound level actually reaching the ear, a hearing conservation program is required. This is to be applied to all employees whose work brings them either steadily or infrequently into areas where sound levels exceed 90dBA.

COMPLIANCE PLAN

Whenever a noise survey shows noise levels in excess of those listed below, steps necessary to reduce the noise exposure to or below those levels shall be ascertained and a detailed plan, with completion dates for individual steps, shall be prepared. Following the original survey which shows existence of overexposures, the steps in a typical compliance plan might include the following:

1. A detailed survey of sound levels and sound spectra to determine the sources of excessive sound levels.
2. Initiation of engineering studies to determine methods for reducing sound levels at their sources.
3. Planning and initiation of feasible administrative controls, such as modifying production schedules to divide noisy jobs among enough people to bring each below the permissible limit or spreading part-time noisy operations.
4. Initial audiograms for personnel excessively exposed.
5. Installation of a personal protective equipment program.
6. Follow-up audiograms at appropriate intervals to assess effectiveness of the personal protective equipment program and administrative controls.
7. Installation of engineering controls, or process changes to reduce noises at their source.
8. Repeated noise surveys to measure the effectiveness of the engineering changes.

When the compliance plan involves long term enginering projects, it may be revised from time to time as conditions change. The orderly completion on schedule of the various phases of the compliance plan, together with other components of the hearing conservation program, will be considered compliance with the regulation.

COMMENTS ON HEARING CONSERVATION

At the time of the writing of this handbook, OSHA had decided that audiometric testing of employees exposed either steadily or infrequently to noise levels above 90dBA was not mandatory. The reason behind the decision was that the standards, as such, did not contain that exact specification in their terminology.

On the other hand, the Walsh-Healey Public Contracts Act, adopted by Secretarial decision as part of OSHA-1970, specified very plainly, simply and clearly the entire approach on hearing conservation, including mandatory audiometric testing. Very plainly, also, as any OSHA or state staff member, from top official to compliance officer knows, there can be no "effective" hearing conservation program without initial and regular follow-up audiometric testing of employees so exposed.

The hearing conservation standard, as published on page 10518, section 1910.95 of the Federal Register, May 29, 1971, Occupational Noise Standard, paragraph b3 specifies: "In all cases where the sound levels exceed the values shown herein (see below) a continuing *effective* hearing conservation program shall be administered." That *is* mandatory and *in* the standards.

Table G-16, Permissible Noise Exposures

Duration per day hours	Sound level dBA slow response
8	90
6	92
4	95
3	97
2	100
1.5	102
1	105
0.5	110
0.25 or less	115

The Act also states that "exposure to impact noise should not exceed 140 dB peak sound pressure level." This sets the upper limit of sound level to which a person should be exposed, regardless of the brevity of the exposure.

In contrast with the 115dBA upper limit for steady noise, the higher intensity for impact noise is permissible because the noise impulse resulting from impacts, like hammer blows or explosive processes, is past the ear before it has time to react fully. Impact noise levels are to be measured only with an impact meter or oscilloscope.

The key word in this critical part of the hearing conservation program approach is "effective." "Effective" as defined in the U. S. Department of Labor's Bulletin 334 "Occupational Noise Standards Guidelines," means that "exposed employees will not suffer continuing deterioration of hearing acuity because of their exposure, but that incipient loss of hearing will be detected, and necessary steps taken to prevent further deterioration before serious hearing loss has occurred." This is impossible without initial and regular audiometric testing.

It is unfortunate that the issue has to be begged at all, but what about the specific terminology of the Act itself, as adopted by Congress, and signed by the President? "By providing medical criteria which will insure, insofar as practicable, that no employee will suffer diminished health, functional capacity or life expectancy as a result of his work exposure." Whether mandatory audiometric testing finds its way into the specific terminology remains to be seen, and in some ways really makes no difference.

Two interesting possibilities, already inherent and implied, occur with the terminology of the standards, as they are without mandatory audiometric testing as part of the specific wording. Suppose, for instance, that an employee or employee representative requests an inspection on the basis that a hearing conservation program is ineffective without audiometric testing? Or suppose he questions the Secretary's decision as being "harmful" or "capricious"?

Aside from any law, Federal or state, there is another very practical basis and consideration on audiometric testing. The following example illustrates this.

In 1956 the pulp and paper industry in the Pacific Northwest initiated a program to conduct audiometric tests of all employees in an admittedly noisy business. At the outset it was difficult to convince management of the desirability of this because of the fear of possible repercussions from employees if they were informed of the results.

The program's basis was very practical. Compensation was available for partial or total hearing loss or impairment. The need for determining the employer's responsibilty was essential. In this way the hearing loss or impairment for which an employer was responsible could be determined. If he had conducted an initial pre-hire test which identified the degree of hearing loss at that point and no further hearing loss occurred thereafter because of a hearing conservation program that enforced the wearing of individual hearing protection devices, the employer would not be responsible for any of the loss.

The need for a base test was obvious, for otherwise how could an employer institute and enforce the use of personal protective hearing devices without proof that they were needed? How could he keep an employee from suffering further deterioration in hearing ability unless he could establish a base level and then enforce the use of protective hearing devices?

Surprisingly, employee reaction to the program was positive rather than negative, as had been rather apprehensively anticipated. Employees were appreciative. They rather expected, in a high level noise area and job, to suffer some hearing loss and were not recriminatory. They were more amenable to the use of protective hearing devices, and there were very few suits.

The wood products industry, though certainly not universally so, followed the lead of one of its component members, and for some years, prior to passage of any federal or state legislation, used audiometric tests for employees regularly in high level noise areas.

The state of Oregon, in its 1972 legislation, adopted a measure requiring initial and annual audiometric testing of all employees exposed to noise levels 90dBA and above. It also passed stringent requirements relating to time off the job prior to testing: 12 hours for 90-94.99dBA, 14 hours for 95-99.99dBA, and 16 hours for 100-plus dBA.

Other states will probably duplicate Oregon's legislation to some degree. This undoubtedly emanates directly and originally from the Williams-Steiger Occupational Safety and Health Act of 1970 and its clearly indicated Congressional intent, even though the Occupational Safety and Health Administration was cautiously ambivalent when this was written.

There are many reasons, then, for initiating a hearing conservation program as guidelined in Bulletin 334. The first is most important for the employer genuinely concerned about his workers.
1. Job-related, noise-induced hearing loss or impairment is, almost without exception, irreversible and permanent.
2. The probability is that both federal and state laws and standards will make audiometric testing mandatory.

3. States that presently have workmen's compensation laws make compensability awards on partial or total hearing loss.
4. Employers need to know what part of the hearing loss, if any, can be charged to them, a vital reason for an initial base level hearing acuity test.
5. Employers need to know if the employee's hearing loss or impairment stays static, with the enforced use of protective hearing devices, or deteriorates. Follow-up regular audiometric testing will reveal this.
6. Employers need to be able to post employees on hearing loss or impairment problems so that the employee can take whatever medical action is necessary, advisable or desired.
7. Employees need a good reason for wearing protective hearing devices.
8. Either employers or unions can initiate audiometric testing as part of a hearing conservation program.
9. As touchy as it is, and as infrequently as it has been employed to date, the possibility of suit cannot be overlooked.

Once the testing has been instituted, employees should be notified of the results of the individual audiometric test. It is not necessary or required that the employee see the test card or audiometric readout. As a matter of fact, interpreting it can be done only by either a qualified, professional audiologist or a physician. But the employee should have the results interpreted for him.

As a matter of further and very important fact, while the audiometric testing can be done by qualified technicians, and while the read-out *can* be done by an audiologist, the Act specifies that it shall be done under the direction of a physician. Also, to be medically and legally acceptable, all audiometric readouts must be signed by a physician, not an audiologist.

Several form letters of notification, which were employed and used successfully for several years by the Weyerhaueser Company in its hearing conservation program, are included here. These can be adapted to meet your needs whether you are conducting your hearing conservation program in-company or having it done by some outside agency.

LETTER TO EMPLOYEE REPORTING HEARING TEST RESULTS

Company: Summit Timber Company
Location: Darrington Washington
Date: January 13, 1972
From: (Company Official)
Subject: Periodic Hearing Tests
To: George A. Armstrong (Individual Tested)

Your initial hearing test on, January 12, shows that:
(date)

(X) Your hearing is essentially normal in all the tones tested.
( ) You show evidence of loss in the higher tones.
( ) You show evidence of loss in the lower tones.
( ) You show evidence of loss in both the higher and lower tones.
While the Occupational Safety and Health Act of 1970 requires that you wear acceptable protective hearing devices continuously on the job, we recommend it even more basically to you as a necessary an practical health measure, to, (1) prevent hearing loss, or, (2 prevent any further hearing loss.
Thank you.
Physician,
  Charles M. McGill,
  M.D., M.P.H.

ANNUAL AUDIOGRAM FORM LETTER

Company
Location
Date
From
Subject
To

Your periodic hearing test on, _____, shows that there has been no change in your hearing since your last test on, _____.
Your hearing is essentially normal in all the tones tested.
It is recommended that you continue to wear ear plugs or muffs on the job continuously, to preserve your normal hearing.

  Signature
  Physician,
  Company Official

ANNUAL AUDIOGRAM FORM LETTER

Company
Location
Date
From
Subject
To

Your periodic hearing test on, _____, shows that there has been no change in your last test on, _____. However, there is some evidence of some prior loss in the higher tones which still persists.

It is recommended that you continue to wear ear plugs or muffs on the job continuously to prevent further hearing loss. Continued protection will prevent your high tone loss from progressing to the lower tones, which are the important ones for the understanding of the human voice in conversation.

If there are any questions about these results and/or recommendations, please check with (Company department) for clarification.

   Signature
   Physician,
   Company Official

ANNUAL AUDIOGRAM FORM LETTER

Company
Location
Date
From
Subject
To

Your periodic hearing test on, _____, shows that there has been no change in your hearing since your last test on, _____ However, there is some evidence of some prior loss in both the higher and lower tones which still persists.

It is recommended that you continue to wear ear plugs or muffs on the job continuously to prevent further hearing loss. Hearing in the lower tones must be preserved to adequately understand the human voice in conversation.

If there are any questions about these results and/or recommendations, please check with (Company department) for clarification.

   Signature
   Physician,
   Company Official

ANNUAL AUDIOGRAM FORM LETTER
Company
Location
Date
From
Subject
To

Your periodic hearing test on, _____, shows a significant change since your last test on, _____. Since then you have lost some hearing in, _____; _____; _____.
                 Low tones   high tones   low and high tones

It is recommended that you:
- ( ) Wear ear plugs or muffs continuously on the job to prevent further hearing loss.
- ( ) Return for a hearing check in _____ months instead of the usual interval. Call the (Company department) before coming in for the test.
- ( ) Come in to the (Company department) to talk to a doctor about your test.

If there are any questions about these results and/or recommendations, please check with the (Company department) for clarification.

  Signature
  Physician,
  Company Official

CHAPTER EIGHT

Communication and Education

INFORMATION DISSEMINATION

This Act applies to practically every workplace, and everyone in that workplace. Be sure, therefore, that everyone knows what the Act is, what it is all about, how it applies.

This is neither simple nor easy. It is, however, necessary and advisable.

1. Begin with management and supervisory personnel.

A. If at all possible, turn your top executive officer out for at least one informational conference on the Act.

He is going to have a difficult time believing it to begin with—no matter how capably *you* tell it. It helps to hear it in person, first-hand.

B. Get other management personnel out to informational conferences and meetings on the Act. Safety Councils, business and industrial associations, professional organizations, sponsor such meetings. OSHA conducts them. State offices hold them. Go and find out.

C. Arrange a meeting of your own management and supervisory group—preferably with some representative of either the federal or state Act as part of it.

Note: If a compliance officer is part of it, hold the meeting away from your plant or place of business. By terms of his job specification a compliance officer must make an inspection any time he visits your plant.

Call your state department of labor or corresponding agency. Check with OSHA, area or region headquarters. Check with your local safety council or chapter of the American Society of Safety Engineers.

2. Going to meetings does not always accomplish the objective you may be seeking. We know of some who, after several conferences didn't really grasp the full significance of the Act, some who still don't, and some who never will.

Management—in any form—sometimes does not really believe the Act, its state parallel, that it is here, that it means what it says, that it applies to your business—until a compliance officer shows up to inspect the facility.

Inspection, as such, is certainly one way to learn about it—but it can be costly, irritating, frustrating and really, inexcusable.

So get the "boss" out, for at least one meeting where he hears the word first hand.

Let's re-emphasize the need for somebody to be fully acquainted with the Act, to be charged with its responsibility, and see that management is as fully acquainted with it, by memo, by report, by summary, however it is formally managed—on paper.

3. Begin to get the word out to employees. The Act says, "The employer shall notify the employee of his rights and obligations, and with reasonable particularity of the standards that are applicable, by notice and by other means."

There is a printed poster, "SAFETY AND HEALTH PROTECTION ON THE JOB", the centerpiece in your Record Keeping Requirements brochure, which must be displayed in a prominent place in the establishment to which your employees normally report to work.

That is only a beginning. Don't leave it at that.

4. Conduct, or cause to be conducted, informational meetings on the Act, standards, specifications, requirements, rights and responsibilities, for employees. For *all* employees.

5. Employees are going to need to know about the Act—just as you do, in order to:

A. Help the company in the achievement of its compliance program.

B. Understand their rights *and* responsibilities under the Act and its standards.

C. Understand the company's rights and responsibilities, and problems, in achieving compliance.

D. Know the need and necessity to comply individually in the development of, and adherence to, safe and healthful work habits.

6. Employers, unions or other outside agencies, can accomplish the employee dissemination discussed here. The employer's decision is a very simple one. How important is it for who to do it?

TRAINING AND EDUCATION

In support of your program of information dissemination, you will need to conduct, cause to be conducted, or send personnel to:

1. Specialized courses in general safety and health subjects.

2. Specialized courses with specific relationship to the Act, standards, or portions thereof.

3. Regular meetings on aspects of, and instruction in, the Act, such as:

A. noise abatement and hearing conservation program.

B. use of personal protective equipment devices.

C. first aid.

4. Specialized courses in leadership, management techniques, motivation and communicating techniques, public relations and information.

5. Meetings, conferences, seminars, congresses on safety and health, for a broader exposure to the problem as such, and to get acquainted with the ways, means, methods employed by others in their approach to the problem.

6. Enrollment in those courses at the vocational school, junior college or university level, with or without credit, that will help your personnel achieve a more effective safety and health program.

PUBLIC RELATIONS—INFORMATION

You would do well to use your safety and health program as a valid means of positive, meaningful and valuable contact with your community.

1. Make any noteworthy event or function, related to the Act, the

subject of regular news coverage for your house organ, newsletter, or simple memo to all personnel.

2. Employ the same tactic with your local newspaper(s), radio and TV station(s).

3. Seek opportunities to speak, make presentations—at service clubs, chambers of commerce, professional societies, trade and industrial associations, PTA's, and other community meeting places and forums.

4. See that those in your company with real expertise on the Act are participants as panel members, principal speakers, at safety and health conferences—and notify your local news media of that participation.

5. Use your safety and health program—or an interesting aspect of it—for an advertisement, or a series of ads.

Remember that what makes news is that it is:
(1) Newsworthy.
(2) Interesting.
(3) Different.
(4) Unusual.
(5) Unique.

Remember that what makes the wastebasket is that it is:
1. Too commercial—like an advertisement for the company.

You don't have to be an accomplished writer or reporter to get a story in the news media. They will *re-write whatever* you turn in anyhow. Just:

1. Compile the facts on a who, what, when, where, why and how basis.
2. Typewrite the facts on a clean sheet of 8½ x 11 paper.
3. Double space the lines.
4. Do an original (no carbons, Xeroxes or duplicates, please).
5. Put the date, your name and telephone number where you can be reached, at the top of the release.
6. Fold the release in half—typed copy out.
7. Mail it, or take it into the city desk, the news editor of a radio or TV station.
8. Sometimes you can solve the entire problem by calling the city desk, or the news editor, who may assign you a reporter to do, or take, the story over the phone.
9. Be sure all names are complete, accurate and spelled right.

Put safety and health stories actively to work in your house organ, if you have one. With pictures.

Include such things as:

(1) Broad corporate policy decisions on major decisions related to the Act.
(2) Initiation of special programming. Hearing conservation, for example. Special training of personnel. Enrollment in safety and health classes by company personnel.
(3) Engineering changes or installation of new equipment to cope with standards requirements of the Act.
(4) Awards or incentive programs for personnel.

Write and send articles to business, industrial and trade associations.

Send notices of any speeches, presentations made by company personnel, to professional society and trade association publications.

Solicit opportunities for acknowledged specialists in your company to appear on panels, programs, presentations.

MOTIVATIONAL AND COMMUNICATING TECHNIQUES
(How to Get the Job Done)

The Act provides the essential, basic and necessary legal framework for getting the job done, but doing it is still up to us, an individual elective employer and employee.

Achieving the real objectives of the Act—the saving of lives, the reduction in numbers, severity and cost of job-related injuries and illnesses—is almost totally dependent upon, "saying it so they hear it."

That's why the information dissemination, training and education, and public relations and information processes mentioned previously are so important.

Here are some cardinal guidelines for communication, one to one, one to several and several to several, to put to work.

1. The basic purposes of *all* communication is—to learn who we are, what we are, what we sound like, what we need to do to survive.
2. The secondary purpose of *all* communication is—to move others to some kind of action. (Belief, acceptance of an idea, doing something about it).

Consider the following precepts, as bald and as unqualified as some may sound.

1. Man's principal function on earth is—survival. Learning what to do, what not to do, how to do it,—to survive.
2. Man is *not* born equal. He is unequal in many, many ways. His "equalness," if you prefer to interpret it that way, is in his very "unequalness." His differences, one from another. His unique-

ness. His sole and exclusive ownership of a set of fingerprints unduplicated anyplace else in the world, or in time.

3. Man is *not* seeking "equality" as such. He does *not* want to be the "equal" of somebody else. He is rather seeking the real inalienable right of mankind—the right to be his own man, the miraculously *different* and *unique* individual he was created to be. He is seeking the right to fulfill that individuality to the best of his ability—and the acknowledgement by others of that simple truth, image and personality, within those limitations set by society for its group as well as its individual survival.

4. Man looks for *acknowledgement* from his peers and others of his *membership in the human race.* He wants no handicaps in *that* equality, but he does not want to be the equal of somebody. He wants to excel by means of his own personal, individual, and in a sense, "unequal" talents.

5. Man, therefore, wants to, deserves to, be treated in that context of equality as a *fellow human being,* but *individually* as a person.

6. He also, therefore, will not be, does not want to be: looked down on, tolerated or put up with; treated as an inferior; put upon or dealt with unfairly or dishonestly; the object of unwarranted disapproval or criticism; considered stupid, unintelligent, incapable, inadequate; set apart, segmented, classified, alienated, or even, in many instances, integrated.

7. He is basically motivated by "self interest"—*however* you happen to define that—practically, philosophically, psychologically, sociologically and physiologically.

If it does something for him, he will listen to it. *He may,* in fact, even *hear you the way you thought you said it—and do it.*

Man, regardless of how sophisticated or knowledgeable, intellectual or learned, how far he has progressed up the ladder of success, however you may define him, is a survivor, whose every talent and capability is directed toward that objective.

Starting from that stringent, but nonetheless realistic, objective and practical ground zero from the standpoint of *your* survival as a would-be effective communicator, keep the following dynamics in mind.

1. All really good communicators are apprehensive, nervous, or just plain scared. That includes the biggest and the best of them, the most successful pros in the business.

Plainly and simply, if you are not, you are certain to bomb out with your audience.

That trembling, shaky, uncomfortable unease, is a physiological reaction to a psychological stimulus—adrenalin pumping into your blood stream to help you meet a crisis or emergency.

No crisis for you? No reaction from your body? No response from your audience? *Dullsville!*

2. All really good communicators know their subject forward and backward.

3. All really good communicators organize their information and material with great care and attention to detail.

4. All really good communicators rehearse assiduously and at length. Call it anything else you like—but rehearsal is its name.

5. All really good communicators establish an immediate rapport with their audience—by taking into account some of the precepts we have discussed above; by talking *to* their audience, not down to or at them; by including all of their audience, looking directly at them—in the eye, responding and reacting to them; by working diligently to secure just that objective—to be one with your audience.

6. All really good communicators take some pains ahead of time to find out about the audience, who they are, what they are, likes, dislikes, prejudices, hangups, no-no's.

7. All really good communicators have a "game plan." **Some** speak completely—or seemingly—ad lib, some speak from notes, some memorize, some read. Memorizing and/or reading is not necessarily recommended, but whatever works best for you, do it, and be sure you have some kind of formalized plan of action, continuity or performance.

8. All really good communicators adapt to the situation—remain flexible even within a formalized "plan of action" to be able to adjust to the situation as it develops, or as it may change for some reason or other.

9. All really good communicators manage such things as "inflection," "gestures," "presentation" by:

A. Observing all the precepts we have mentioned

B. Knowing the subject thoroughly and being thoroughly prepared

C. Enthusiasm

Such dynamics as "inflection," "gestures," "presentation," spring naturally, becomingly and emphatically out of the three bases we have noted above.

If they are "practiced," "contrived," or used consciously, they are apt to detract from the presentation.

10. All really good communicators, insofar as is practical and possible, "case the joint" ahead of time. Take the time to take an in-person look at the place of presentation and any problems pre-

sented thereto. Particularly if you are using audio visual aids.
11. All really good communicators develop, employ, and use any audio-visual aid helpful to the cause. Make sure ahead of time they are set up, functioning and ready to go.
12. All really good communicators (even though they may suffer the torture of the damned) love, enjoy and have fun communicating. If *you* don't—forget it.

NOTE

Everything mentioned above applies in some degree to every communicating situation in which you find yourself. Conversation with somebody, command decisioning as a foreman, supervisor, superintendent, manager, director, vice-president, president, or chairman of the board.

Making presentations, speeches, awards, acceptances. Acting, singing, entertaining, performing. *To name a few.*

Perhaps the most difficult of all—all really good communicators are excellent listeners.

EMPLOYEE RELATIONS

Many authorities on the Act and its implications consider it as much an employee relations bill as a safety and health dictum. Its employee relations aspect, effectively implemented, is noteworthy, considerable, and has a practical potential which is almost entirely overlooked.

To begin with, no legislation as such accomplishes a specified objective of itself. It needs people, people working together, cooperating, contributing, participating, seeking a mutually profitable, desirable and achievable goal.

Look at the *real* goals of the Act—although not specified there in the terminology below.
1. Saving life.
2. Reducing the numbers, severity and cost of job related accidents and illnesses.
3. Helping establish the security and well-being of those who work with and for us, who help get the job done.
4. Profit-making.

Profit-making?

That's right. Business and industry, with minimal exception, have never tied the cost of accidents to profit-making—or profit-losing.

Production, production increase, more efficient production, uninterrupted production, has been the traditional answer—not preventive protectivity.

If a preventable accident occurs involving, for example, $100,000 in direct costs, to an employee, however that spells out or adds up —it has to be $100,000 in lost profit.

If insurance rates for a company are increased because of an accelerating accident frequency ratio, that could be reduced or at least held static by effective loss control, accident prevention, safety and health programming measures—the increase in rate has to add up to lost profit.

Traditionally, safety has gone begging. Perhaps this is due to the fact that production modes, changes, are specific, tangible, observable and assessable. Accident prevention, loss control, safety management, is seemingly intangible, hard to fix and assess in numbers and results, and requiring in many ways faith in things unseen, demonstrating little in terms of cold, logical and recognizable return.

If we come back to those four goals again, we will look for a long time before we find anything more in the favor of both the employer and employee as a practical meeting ground for partnering-up accomplishments, or anything more in their mutual self interest.

Many employer or management people are apprehensive, if indeed not downright bitter about the "employee rights" aspect of the Act. Justifiable or not, one thing for certain is—that it is not realistic.

It also overlooks a vitally important plus that employees rights within the Act specification represents. It is only a negative if viewed that way by management—and if that is the overt or covert attitude of management, it not only can be, but is a real negative.

If, for example, management positions itself against the Act, demonstrates its opposition to it in whatever observable form, it is not only readily evident *to* employees, but carries along with it an also obvious message.

1. Sorry. Your life and limb and health don't really mean that much to us.
2. We're really not interested in you.
3. After money you come first.

There are all kinds of legitimate, valid reasons for disagreeing with the mechanics and the niggling detail of the Act—but obvious and expressed disagreement with its basic principles is like arguing against mother love.

Viewed positively and practically, as an excellent employee relations opportunity, what are some of the steps that could be taken?

1. We have already suggested a number of them in the preceding pages.
2. Develop a position paper that clearly defines company philosophy, attitude, policy and programming approach.
3. Be sure that all management and supervisory personnel is thoroughly acquainted with that position and adheres to it. (No gripers or cry-babies please, particularly in leadership positions.)
4. Let your employees—all employees—know what that position is.
5. State the problem the Act and compliance with it poses the company, employer *and* employee.
6. Secure involvement and participation in solutions to the problem from as broad a representative group, from total company structure, as you can possibly manage.
7. Solicit answers to the problem from total company personnel.
8. Act on these solutions and answers immediately, or as soon as possible. If action must be delayed, be sure the reasons for it are specified, explained and understood.
9. Institute some kind of awards program.
10. Actively foster, support and participate in:
A. Information dissemination to total company structure.
B. Education and training programs.
C. Membership in professional safety and health societies, organizations, councils, committees.
D. Meetings, conferences, seminars, congresses on safety and health.
11. Trust and include employees as equal, responsible, interested, contributing, participating and vitally important members of the company production team.

Since the first employer rewarded or paid the first employee for the work performed, a relationship was forged as unbreakable as the simple need to survive for both. Different in degree, but nonetheless the same basically, and made possible *by* that employment.

That simple relationship, a kind of silver cord, employer to employee does exist. Employees, generally speaking, are loyal to their employer. Their first allegiance on the job is to that employer. There is therefore a working relationship—employer to employee—that can be interrupted, bargained and bartered, but not severed as long as the worker works for the man who hired him and who pays his wages.

Employers have a lot more going for them in this relationship than they are inclined to think.

Employee rights, as specified in the Act, have faced the employer, perhaps for the first time, with some specific decisions he has not had to make prior to the passage of the Act.

Since real compliance with the Act, the achievement of its goals and objectives, in its end result is totally dependent upon people, how they think, respond, react, cooperate and participate, it behooves the employer from a practical standpoint alone to take every means possible to make employee rights work *for* him.

Employees are quick to learn the company attitude, and equally quick to respond in duplicate.

The Act, believe it or not, is a remarkable employee relations opportunity for the employer, and member-relations opportunity for the unions.

As a perhaps noteworthy comment on this aspect of the Williams-Steiger Occupational Safety and Health Act of 1970, listen.

At an annual industrial safety conference I stepped out of a session in which a management representative made a rather large point, in front of an equally large audience, of the fact that management had vigorously opposed the Act, from its outset.

Standing in a small conversational group in the corridors of this meeting area we were joined by, and introduced to, a man who turned out to be a rather high ranking labor official. He was irked by what he had just heard in that same session, as indeed was I.

After some pros and cons on the subject, the following statement was made:

"If we're not careful," he said, **"the worker is going to make an end run around both of us—management** *and* **labor."**

CHAPTER NINE
Index to Part 1910

OCCUPATIONAL SAFETY AND HEALTH STANDARDS
FEDERAL REGISTER VOL. 36, NUMBER 105,
SATURDAY, MAY 29, 1971

Heading	Subpart	Section	Page
General	A		10467
Purpose and Scope		1910.1	10467
Definitions		1910.2	10467
Petitions for the issuance, amendment, or repeal of a standard		1910.3	10468
Amendments to this part		1910.4	10468
Applicability of standards		1910.5	10468
Incorporation by reference		1910.6	10468
Adoption and Extension of Established Federal Standards	B		10468
Scope and Purpose		1910.11	10468
Construction Work		1910.12	10469
Ship Repairing		1910.13	10469
Ship Building		1910.14	10469
Shipbreaking		1910.15	10469
Longshoring		1910.16	10469
(Reserved)	C		10469
Walking—Working Surfaces	D		10469
Definitions		1910.21	10469

Heading	Subpart Section	Page
General Requirements	1910.22	10472
Guarding floor and wall openings and holes	1910.23	10472
Protection for floor openings	1910.23(a)	10472
Protection for wall openings & holes	1910.23(b)	10473
Protection of open-sided floors, platforms and runways	1910.23(c)	10473
Stairway railings and guards	1910.23(d)	10473
Railing, toe boards and cover specifications	1910.23(e)	10473
Fixed industrial stairs	1910.24	10474
Application of requirements	1910.24(a)	10474
Where fixed stairs are required	1910.24(b)	10474
Stair strength	1910.24(c)	10474
Stair width	1910.24(d)	10474
Angle of stairway rise	1910.24(e)	10474
Stair treads	1910.24(f)	10474
Length of stairways	1910.24(g)	10475
Railings and handrails	1910.24(h)	10475
Vertical clearance	1910.24(i)	10475
Open risers	1910.24(j)	10475
General	1910.24(k)	10475
Portable wood ladders	1910.25	10475
Application of requirements	1910.25(a)	10475
Materials	1910.25(b)	10475
Construction requirements	1910.25(c)	10475
Care and use of ladders	1910.25(d)	10480
Portable metal ladders	1910.26	10481
Requirements	1910.26(a)	10481
Testing	1910.26(b)	10482
Care, Maintenance and Use of Ladders	1910.26(c)	10482
Fixed ladders	1910.27	10483
Design requirements	1910.27(a)	10483
Specific features	1910.27(b)	10483
Clearance	1910.27(c)	10483
Special requirements	1910.27(d)	10484
Pitch	1910.27(e)	10485
Maintenance	1910.27(f)	10486
Safety requirements for scaffolding	1910.28	10486
General requirements for all scaffolds	1910.28(a)	10486

OSHA HANDBOOK 65

Heading	Subpart Section	Page
General requirements for wood pole scaffolds	1910.28(b)	10486
Tube and coupler scaffolds	1910.28(c)	10488
Tubular welded frame scaffolds	1910.28(d)	10488
Outrigger scaffolds	1910.28(e)	10489
Masons' adjustable multiple-point suspension scaffolds	1910.28(f)	10489
Two-point suspension scaffolds (swinging scaffolds)	1910.28(g)	10489
Stone setters' adjustable multiple-point suspension scaffolds	1910.28(h)	10490
Single-point adjustable suspension scaffolds	1910.28(i)	10490
Boatswain's chairs	1910.28(j)	10490
Carpenters' bracket scaffolds	1910.28(k)	10491
Bricklayers' square scaffolds	1910.28(l)	10491
Horse scaffolds	1910.28(m)	10491
Needle beam scaffolds	1910.28(n)	10491
Plasters', decorators' and large area scaffolds	1910.28(o)	10491
Interior hung scaffolds	1910.28(p)	10491
Ladder-jack scaffolds	1910.28(q)	10492
Window-jack scaffolds	1910.28(r)	10492
Roofing brackets	1910.28(s)	10492
Crawling boards or chicken ladders	1910.28(t)	10492
Float or ship scaffolds	1910.28(u)	10492
Manually propelled mobile ladder stands and scaffolds (towers)	1910.29	10492
General requirements	1910.29(a)	10492
Mobile tubular welded frame scaffolds	1910.29(b)	10493
Mobile tubular welded sectional folding scaffolds	1910.29(c)	10493
Mobile tube and coupler scaffolds	1910.29(d)	10493
Mobile work platforms	1910.29(e)	10493
Mobile ladder stands	1910.29(f)	10493
Other working surfaces	1910.30	10494
Dockboards	1910.30(a)	10494
Forging machine area	1910.30(b)	10494
Sources of standards	1910.31	10494
Standards organizations	1910.32	10494

Heading	Subpart	Section	Page
Means of Egress	E		10494
Definitions		1910.35	10494
General requirements		1910.36	10494
Application		1910.36(a)	10494
Fundamental requirements		1910.36(b)	10494
Protection of employees exposed by construction and repair operations		1910.36(c)	10495
Maintenance		1910.36(d)	10495
Means of egress, general		1910.37	10495
Permissible exit components		1910.37(a)	10495
Protective enclosure of exits		1910.37(b)	10495
Width and capacity of means of egress		1910.37(c)	10495
Egress capacity and occupant load		1910.37(d)	10495
Arrangement of exits		1910.37(e)	10495
Access to exits		1910.37(f)	10495
Exterior ways of exit access		1910.37(g)	10495
Discharge from exits		1910.37(h)	10496
Headroom		1910.37(i)	10496
Changes in elevation		1910.37(j)	10496
Maintenance and workmanship		1910.37(k)	10496
Furnishings and decorations		1910.37(l)	10496
Automatic sprinkler systems		1910.37(m)	10496
Alarm and fire detection systems		1910.37(n)	10496
Fire retardant paints		1910.37(o)	10496
Recognition of means of egress		1910.37(p)	10496
Exit marking		1910.37(q)	10496
Specific means of egress requirements by occupancy (Reserved)		1910.38	10496
Sources of standards		1910.39	10496
Standards organizations		1910.40	10496
Powered Platforms, Manlifts and Vehicle-Mounted Work Platforms	F		10496
Power platforms for exterior building maintenance		1910.66	10496
Definitions applicable to this section		1910.66(a)	10496
General requirements		1910.66(b)	10497
Type F powered platforms		1910.66(c)	10497
Type T powered platforms		1910.66(d)	10500
Vehicle-mounted elevating and rotating work platforms		1910.67	10501

OSHA HANDBOOK 67

Heading	Subpart Section	Page
Definitions applicable to this section	1910.67(a)	10501
General requirements	1910.67(b)	10501
Man lifts	1910.68	10501
Definitions applicable to this section	1910.68(a)	10501
General requirements	1910.68(b)	10501
Mechanical requirements	1910.68(c)	10502
Operating rules	1910.68(d)	10503
Periodic inspection	1910.68(e)	10503
Sources of standards	1910.69	10503
Standards organizations	1910.70	10503
Occupational Health and Environmental Control	G	10503
Air contaminants (Gases, vapors, fumes, dust, and mists)	1910.93	10503
Ventilation	1910.94	10506
Abrasive blasting	1910.94(a)	10506
Grinding, polishing and buffing operations	1910.94(b)	10507
Spray-finishing operations	1910.94(c)	10512
Open surface tanks	1910.94(d)	10515
Occupational noise exposure		
Ionizing radiation	1910.95	10518
Definitions applicable to this section	1910.96	10518
Exposure of individuals to radiation in restricted areas	1910.96(a) 1910.96(b)	10518 10518
Exposure to airborne radioative material	1910.96(c)	10519
Precautionary procedures and personal monitoring	1910.96(d)	10519
Caution signs, labels and signals	1910.96(e)	10519
Immediate evacuation warning signal	1910.96(f)	10520
Exceptions from posting requirements	1910.96(g)	10521
Exceptions for radioactive materials packaged for shipment	1910.96(h)	10521
Instruction of personnel posting	1910.96(i)	10521
Storage of radioactive materials	1910.96(j)	10521
Waste disposal	1910.96(k)	10521
Notification of incidents	1910.96(l)	10521
Reports of overexposure and excessive levels and concentrations	1910.96(m)	10521

68   OSHA HANDBOOK

Heading	Subpart Section	Page
Records	1910.96(n)	10521
Disclosure to former employee of individual employee's record	1910.96(o)	10522
Atomic Energy Commission licenses	1910.96(p)	10522
Radiation standards for mining	1910.96(r)	10522
Nonionizing radiation	1910.97	10522
Electromagnetic radiation	1910.97(a)	10522
Additional delay in effective date	1910.98	10523
Sources of standards	1910.99	10523
Standards organizations	1910.100	10523
Hazardous Materials	H	10523
Compressed gases (general requirements)	1910.101	10523
Inspection of compressed gas cylinders	1910.101(a)	10523
Compressed gases	1910.101(b)	10524
Safety relief devices for compressed gas containers	1910.101(c)	10524
Acetylene	1910.102	10524
Cylinders	1910.102(a)	10524
Piped systems	1910.102(b)	10524
Generators and filling cylinders	1910.102(c)	10524
Hydrogen	1910.103	10524
General	1910.103(a)	10524
Gaseous hydrogen systems	1910.103(b)	10524
Liquified hydrogen systems	1910.103(c)	10525
Oxygen	1910.104	10528
General	1910.104(a)	10528
Bulk oxygen systems	1910.104(b)	10528
Nitrous oxide	1910.105	10529
Flammable and combustible liquids	1910.106	10529
Definitions	1910.106(a)	10529
Tank storage	1910.106(b)	10530
Piping valves and fittings	1910.106(c)	10536
Container and portable tank storage	1910.106(d)	10537
Industrial plants	1910.106(e)	10539
Bulk plants	1910.106(f)	10540
Service stations	1910.106(g)	10543
Processing plants	1910.106(h)	10545
Refineries, chemical plants and distilleries	1910.106(i)	10546

Heading	Subpart Section	Page
Spray finishing using flammable and combustible materials	1910.107	10546
Definitions applicable to this section	1910.107(a)	10546
Spray booths	1910.107(b)	10547
Electrical and other sources of ignition	1910.107(c)	10547
Ventilation	1910.107(d)	10548
Flammable and combustible liquids—storage and handling	1910.107(e)	10548
Protection	1910.107(f)	10549
Operations and maintenance	1910.107(g)	10549
Fixed electrostatic apparatus	1910.107(h)	10549
Electrostatic hand spraying equipment	1910.107(i)	10550
Drying, curing or fusion apparatus	1910.107(j)	10550
Automobile undercoating in garages	1910.107(k)	10550
Powder coating	1910.107(l)	10550
Organic peroxides and dual component coatings	1910.107(m)	10551
Dip tanks containing flammable or combustible liquids	1910.108	10551
Definitions applicable to this section	1910.108(a)	10551
Ventilation	1910.108(b)	10551
Construction of dip tanks	1910.108(c)	10551
Liquids used in dip tanks, storage and handling	1910.108(d)	10551
Electrical and other sources of ignition	1910.108(e)	10552
Operations and maintenance	1910.108(f)	10552
Extinguishment	1910.108(g)	10552
Special dip tank applications	1910.108(h)	10552
Explosives and blasting agents	1910.109	10553
Definitions applicable to this section	1910.109(a)	10553
Miscellaneous provisions	1910.109(b)	10554
Storage of explosives	1910.109(c)	10554
Transportation of explosives	1910.109(d)	10556
Use of explosives and blasting agents	1910.109(e)	10557
Explosives at piers, railway stations and cars or vessels not otherwise specified in this standard	1910.109(f)	10557
Blasting agents	1910.109(g)	10557
Water gel (Slurry) explosives and blasting agents	1910.109(h)	10560

Heading	Subpart Section	Page
Storage of ammonium nitrate	1910.109(i)	10561
Storage and handling of liquified petroleum gases	1910.110	10562
Definitions applicable to this section	1910.110(a)	10562
Basic rules	1910.110(b)	10563
Cylinder systems	1910.110(c)	10571
Systems utilizing containers other than DOT containers	1910.110(d)	10573
Liquified petroleum gas as a motor fuel	1910.110(e)	10576
Storage of containers awaiting use or resale	1910.110(f)	10578
LP-Gas system installations on commercial vehicles	1910.110(g)	10578
Liquified petroleum gas service stations	1910.110(h)	10581
Storage and handling of anhydrous ammonia	1910.111	10583
General	1910.111(a)	10583
Basic rules	1910.111(b)	10583
Systems utilizing stationary, nonrefrigerated storage containers	1910.111(c)	10586
Refrigerated storage systems	1910.111(d)	10587
Systems utilizing portable DOT containers	1910.111(e)	10588
Tank motor vehicles for the transportation of ammonia	1910.111(f)	10588
Systems mounted on farm vehicles other than for the application of ammonia	1910.111(g)	10589
Additional delay in effective date	1910.114	10589
Sources of standards	1910.115	10589
Standards organizations	1910.116	10590
Personal Protective Equipment	I	10590
General requirements	1910.132	10590
Application	1910.132(a)	10590
Employee-owned equipment	1910.132(b)	10590
Design	1910.132(c)	10590
Eye and face protection	1910.133	10590

OSHA HANDBOOK 71

Heading	Subpart Section	Page
Respiratory protection		
Permissible practice	1910.134	10590
Requirements for a minimal acceptable program	1910.134(a)	10590
	1910.134(b)	10590
Selection of respirators	1910.134(c)	10590
Air quality	1910.134(d)	10590
Use of respirators	1910.134(e)	10591
Maintenance and care of respirators	1910.134(f)	10591
Identification of gas mask canisters	1910.134(g)	10592
Occupational head protection	1910.135	10592
Occupational foot protection	1910.136	10592
Electrical protective devices	1910.137	10592
Additional delay in effective date	1910.138	10592
Sources of standards	1910.139	10593
Standards organizations	1910.140	10593
General Environmental Controls	**J**	10593
Sanitation	1910.141	10593
General requirements	1910.141(a)	10593
Water supply	1910.141(b)	10593
Toilet facilities	1910.141(c)	10593
Washing facilities	1910.141(d)	10594
Change rooms	1910.141(e)	10594
Retiring rooms for women	1910.141(f)	10594
Lunchrooms	1910.141(g)	10594
Food handling	1910.141(h)	10594
Temporary labor camps	1910.142	10594
Nonwater carriage disposal systems	1910.143	10595
Acceptable industrial disposal systems	1910.143(a)	10595
Privy specifications	1910.143(b)	10596
Chemical toilet specifications	1910.143(c)	10596
Seepage pit construction	1910.143(d)	10596
Combustion toilet	1910.143(e)	10596
Recirculating toilet specifications	1910.143(f)	10597
Portable toilet construction	1910.143(g)	10597
Safety color code for marking physical hazards	1910.144	10597
Specifications for accident prevention signs and tags	1910.145	10597
Scope	1910.145(a)	10597

Heading	Subpart Section	Page
Definitions	1910.145(b)	10597
Classification of signs according to use	1910.145(c)	10597
Sign design and colors	1910.145(d)	10597
Sign wordings	1910.145(e)	10599
Accident prevention tags	1910.145(f)	10600
Sources of standards	1910.147	10601
Standards organizations	1910.148	10601
Medical and First Aid	K	10601
Fire Protection	L	10601
Definitions applicable to this subpart	1910.156	10601
Portable fire extinguishers	1910.157	10602
General requirements	1910.157(a)	10602
Selection of extinguishers	1910.157(b)	10602
Distribution of portable fire extinguishers	1910.157(c)	10603
Inspection, maintenance and hydrostatic tests	1910.157(d)	10603
Standpipe and hose systems	1910.158	10604
General requirements	1910.158(a)	10604
Hose outlets	1910.158(b)	10604
Water supplies	1910.158(c)	10604
Tests and maintenance	1910.158(d)	10605
Automatic sprinkler systems	1910.159	10605
General requirements	1910.159(a)	10605
Fire department connections	1910.159(b)	10605
Sprinkler alarms	1910.159(c)	10605
Maintenance of sprinkler system	1910.159(d)	10605
Sprinkler head clearance	1910.159(e)	10605
Fixed dry chemical extinguishing systems	1910.160	10605
General requirements	1910.160(a)	10605
Alarms and indicators	1910.160(b)	10606
Inspection and maintenance	1910.160(c)	10606
Carbon dioxide extinguishing systems	1910.161	10606
General requirements	1910.161(a)	10606
Inspection and maintenance	1910.161(b)	10606
Other special fixed extinguishing systems (Reserved)	1910.162	10606
Local fire alarm signaling systems	1910.163	10606
General requirements	1910.163(a)	10606
Fire alarm boxes	1910.163(b)	10606
Maintenance	1910.163(c)	10606

OSHA HANDBOOK 73

Heading	Subpart Section	Page
Fire brigades (Reserved)	1910.164	10606
Additional delay in effective date	1910.165	10606
Sources of standards	1910.165(a)	10606
Standards organizations	1910.165(b)	10607
Compressed Gas and Compressed Air Equipment	**M**	10607
Inspection of compressed gas cylinders	1910.166	10607
Definitions	1910.166(a)	10607
General requirements	1910.166(b)	10607
Inspection of low-pressure cylinders exempt from the hydrostatic test including acetylene cylinders	1910.166(c)	10607
Low-pressure cylinders subject to hydrostatic testing	1910.166(d)	10608
High-pressure cylinders	1910.166(e)	10608
Internal inspection	1910.166(f)	10609
Safety relief devices for compressed gas cylinders	1910.167	10609
Definitions	1910.167(a)	10609
General requirements	1910.167(b)	10609
Safety relief devices for cargo and portable tanks storing compressed gases	1910.168	10610
Definitions applicable to this section	1910.168(a)	10610
General requirements	1910.168(b)	10611
Air receivers	1910.169	10611
General requirements	1910.169(a)	10611
Installation and equipment requirements	1910.169(b)	10611
Sources of standards	1910.170	10612
Standards organizations	1910.171	10612
Materials Handling and Storage	**N**	10612
Handling Materials—general	1910.176	10612
Use of mechanical equipment	1910.176(a)	10612
Secure storage	1910.176(b)	10612
Housekeeping	1910.176(c)	10612
Drainage	1910.176(d)	10612
Clearance limits	1910.176(e)	10612
Rolling railroad cars	1910.176(f)	10612
Guarding	1910.176(g)	10612

Heading	Subpart Section	Page
Indoor general storage	1910.177	10612
Definitions applicable to this section	1910.177(a)	10612
General requirements	1910.177(b)	10612
Piling procedures and precautions	1910.177(c)	10613
Fire protection requirements	1910.177(d)	10613
Mechanical handling equipment	1910.177(e)	10613
Building service equipment	1910.177(f)	10613
Smoking	1910.177(g)	10613
Powered industrial trucks	1910.178	10613
General requirements	1910.178(a)	10613
Designations	1910.178(b)	10613
Designated locations	1910.178(c)	10614
Converted industrial trucks	1910.178(d)	10615
Safety guards	1910.178(e)	10615
Fuel handling and storage	1910.178(f)	10615
Changing and charging storage batteries	1910.178(g)	10616
Lighting for operating areas	1910.178(h)	10616
Control of noxious gases and fumes	1910.178(i)	10616
Dockboards (bridge plates)	1910.178(j)	10616
Trucks and railroad cars	1910.178(k)	10616
Operator training	1910.178(l)	10616
Truck operations	1910.178(m)	10616
Traveling	1910.178(n)	10616
Loading	1910.178(o)	10617
Operation of the truck	1910.178(p)	10617
Maintenance of industrial trucks	1910.178(q)	10617
Overhead and gantry cranes	1910.179	10617
Definitions applicable to this section	1910.179(a)	10617
General requirements	1910.179(b)	10618
Cabs	1910.179(c)	10618
Footwalks and ladders	1910.179(d)	10619
Stops, bumpers, rail sweeps, and guards	1910.179(e)	10619
Brakes	1910.179(f)	10619
Electric equipment	1910.179(g)	10620
Hoisting equipment	1910.179(h)	10620
Warning device	1910.179(i)	10621
Inspection	1910.179(j)	10621
Testing	1910.179(k)	10621

Heading	Subpart Section	Page
Maintenance	1910.179(l)	10621
Rope inspection	1910.179(m)	10622
Handling the load	1910.179(n)	10622
Other requirements, general	1910.179(o)	10622
Crawler locomotive and truck cranes	1910.180	10622
Definitions applicable to this section	1910.180(a)	10622
General requirements	1910.180(b)	10623
Load ratings	1910.180(c)	10623
Inspection classification	1910.180(d)	10624
Testing	1910.180(e)	10624
Maintenance procedure	1910.180(f)	10624
Rope inspection	1910.180(g)	10624
Handling the load	1910.180(h)	10625
Other requirements	1910.180(i)	10625
Operating near electric power lines	1910.180(j)	10625
Derricks	1910.181	10626
Definitions applicable to this section	1910.181(a)	10626
General requirements	1910.181(b)	10626
Load ratings	1910.181(c)	10627
Inspection	1910.181(d)	10627
Testing	1910.181(e)	10628
Maintenance	1910.181(f)	10628
Rope inspection	1910.181(g)	10628
Operations of derricks	1910.181(h)	10628
Handling the load	1910.181(i)	10628
Other requirements	1910.181(j)	10628
Additional delay in effective date	1910.182	10629
Sources of standards	1910.183	10629
Standards organizations	1910.184	10629
Machinery and Machine Guarding	O	10629
Definitions	1910.211	10629
General requirements for all machines	1910.212	10632
Machine guarding	1910.212(a)	10632
Anchoring fixed machinery	1910.212(b)	10633
Woodworking machinery requirements	1910.213	10633
Machinery construction, general	1910.213(a)	10633
Machine controls and equipment	1910.213(b)	10633
Handfed ripsaws	1910.213(c)	10633
Handfed crosscut table saws	1910.213(d)	10633
Circular resaws	1910.213(e)	10634

Heading	Subpart Section	Page
Self-feed circular saws	1910.213(f)	10634
Swing cutoff saws	1910.213(g)	10634
Radial saws	1910.213(h)	10634
Band saws and band resaws	1910.213(i)	10634
Jointers	1910.213(j)	10634
Tenoning machines	1910.213(k)	10634
Boring and mortising machines	1910.213(l)	10635
Woodshapers and similar equipment	1910.213(m)	10635
Planing, molding, sticking and matching machines	1910.213(n)	10635
Profile, swing-head lathes and wood head turning machine	1910.213(o)	10635
Sanding machines	1910.213(p)	10635
Veneer cutters and wringers	1910.213(q)	10635
Miscellaneous woodworking machines	1910.213(r)	10635
Inspection and maintenance of woodworking machinery	1910.213(s)	10636
Cooperage machinery	1910.214	10636
Abrasive wheel machinery	1910.215	10637
General requirements	1910.215(a)	10637
Guarding of abrasive wheel machinery	1910.215(b)	10637
Flanges	1910.215(c)	10638
Mounting	1910.215(d)	10639
Mills and calendars in the rubber and plastic industries	1910.216	10642
General requirements	1910.216(a)	10642
Mill safety controls	1910.216(b)	10642
Calendar safety controls	1910.216(c)	10642
Protection by location	1910.216(d)	10642
Trip and emergency switches	1910.216(e)	10642
Stopping limits	1910.216(f)	10642
Alarm	1910.216(g)	10643
Mechanical power presses	1910.217	10643
General requirements	1910.217(a)	10643
Mechanical power press guarding and construction, general	1910.217(b)	10643
Safeguarding the point of operation	1910.217(c)	10644
Design, construction, setting and feeding of dies	1910.217(d)	10645
Inspection, maintenance and modification of presses	1910.217(e)	10645

Heading	Subpart Section	Page
Operation of power presses	1910.217(f)	10645
Forging machines	1910.218	10646
General requirements	1910.218(a)	10646
Hammers, general	1910.218(b)	10646
Presses	1910.218(c)	10646
Power-driven hammers	1910.218(d)	10646
Gravity hammers	1910.218(e)	10647
Forging presses	1910.218(f)	10647
Trimming presses	1910.218(g)	10647
Upsetters	1910.218(h)	10647
Other forging equipment	1910.218(i)	10647
Other forge facility equipment	1910.218(j)	10647
Mechanical power-transmission apparatus	1910.219	10647
General requirements	1910.219(a)	10647
Prime-mover guards	1910.219(b)	10647
Shafting	1910.219(c)	10648
Pulleys	1910.219(d)	10648
Belt, rope, and chain drive	1910.219(e)	10648
Gears, sprockets and chains	1910.219(f)	10649
Guarding friction drives	1910.219(g)	10649
Keys, setscrews and other projections	1910.219(h)	10649
Collars and couplings	1910.219(i)	10649
Bearings and facilities for oiling	1910.219(j)	10649
Guarding of clutches, cutoff couplings and clutch pulleys	1910.219(k)	10649
Belt shifters, clutches, shippers, poles, perches and fasteners	1910.219(l)	10649
Standard guards—general requirements	1910.219(m)	10649
Disk, shield and "U" guards	1910.219(n)	10649
Approved materials	1910.219(o)	10650
Care of equipment	1910.219(p)	10651
Additional delay in effective date	1910.220	10651
Sources of standards	1910.221	10651
Standards organizations	1910.222	10652
Hand and Portable Powered Tools and Other Hand-Held Equipment	P	10652
Definitions	1910.241	10652
Explosive-actuated fastening tool terms	1910.241(a)	10652
Abrasive wheel terms	1910.241(b)	10652

Heading	Subpart Section	Page
Power lawnmower terms	1910.241(c)	10652
Jack terms	1910.241(d)	10653
Hand and portable powered tools and equipment, general	1910.242	10653
General requirements	1910.242(a)	10653
Compressed air used for cleaning	1910.242(b)	10653
Modification and procurement of safety equipment	1910.242(c)	10653
Guarding of portable powered tools	1910.243	10653
Woodworking, portable powered tools	1910.243(a)	10653
Pneumatic powered tools and hose	1910.243(b)	10653
Portable abrasive wheels	1910.243(c)	10653
Explosive-actuated fastening tools	1910.243(d)	10654
Power lawnmowers	1910.243(e)	10655
Other portable tools and equipment	1910.244	10655
Jacks	1910.244(a)	10656
Abrasive blast cleaning nozzles	1910.244(b)	10656
Additional delay in effective date	1910.245	10656
Sources of standards	1910.246	10656
Standards organizations	1910.247	10656
Welding, Cutting, and Brazing	**Q**	10656
Definitions	1910.251	10656
Welding, cutting and brazing	1910.252	10656
Installation and operation of oxygen-fuel gas systems for welding and cutting	1910.252(a)	10656
Application, installation and operation of arc welding and cutting equipment	1910.252(b)	10663
Installation and operation of resistance welding equipment	1910.252(c)	10664
Fire prevention and protection	1910.252(d)	10665
Protection of personnel	1910.252(e)	10666
Health protection and ventilation	1910.252(f)	10667
Industrial applications	1910.252(g)	10668
Sources of standards	1910.253	10668
Standards organizations	1910.254	10668
Special Industries	**R**	10669
Pulp, paper, and paperboard mills	1910.261	10669
Textiles	1910.262	10676
Bakery Equipment	1910.263	10679

Heading	Subpart Section	Page
Laundry machinery and operations	1910.264	10687
Sawmills	1910.265	10689
Pulpwood logging	1910.266	10696
Agricultural operations	1910.267	10699
Electrical	S	10699
Application	1910.308	10699
General	1910.308(a)	10699
Coverage of subpart	1910.308(b)	10699
Definition application to this subpart	1910.308(c)	10699
National Electrical Code	1910.309	10699
General	1910.310	10699
Voltages	1910.310(a)	10699
Conductor gauges	1910.310(b)	10699
Conductors	1910.310(c)	10699
Deteriorating agencies	1910.310(d)	10699
Mechanical execution of work	1910.310(e)	10699
Mounting of equipment	1910.310(f)	10699
Connections to terminals	1910.310(g)	10699
Splicers	1910.310(h)	10700
Working space about electrical equipment (600 volts or less)	1910.310(i)	10700
Guarding of live parts (600 volts or less)	1910.310(j)	10700
Arcing parts	1910.310(k)	10700
Light and power from railway conductors	1910.310(l)	10700
Insulation resistance	1910.310(m)	10700
Marking	1910.310(n)	10700
Identification	1910.310(o)	10700
Overcurrent protection	1910.312	10700
Protection of equipment	1910.312(a)	10700
Interrupting capacity	1910.312(b)	10700
Circuit impedance and other characteristics	1910.312(c)	10700
Location in premises	1910.312(d)	10700
Enclosures for overcurrent devices	1910.312(e)	10700
Arcing or suddenly moving parts	1910.312(f)	10700
Grounding	1910.314	10701
Circuit and system grounding	1910.314(a)	10701
Location of grounding connections	1910.314(b)	10701
Enclosure grounding	1910.314(c)	10701

Heading	Subpart Section	Page
Equipment grounding	1910.314(d)	10701
Methods of grounding	1910.314(e)	10702
Bonding	1910.314(f)	10702
Instrument transformers, relays, etc.	1910.314(g)	10702
Outlet, switch, and junction boxes, and fittings	1910.315	10702
Round boxes	1910.315(a)	10702
Nonmetallic boxes	1910.315(b)	10702
Metallic boxes	1910.315(c)	10702
Damp or wet locations	1910.315(d)	10702
Number of conductors in a box	1910.315(e)	10702
Conductors entering boxes or fittings	1910.315(f)	10703
Unused openings	1910.315(g)	10703
Boxes enclosing flush devices	1910.315(h)	10703
In wall or ceiling	1910.315(i)	10703
Repairing plaster	1910.315(j)	10703
Exposed extensions	1910.315(k)	10703
Supports	1910.315(l)	10703
Depth of outlet boxes for concealed work	1910.315(m)	10703
Covers and canopies	1910.315(n)	10703
Accessibility of junction, pull, and outlet boxes	1910.315(o)	10703
Flexible cords and cables	1910.316	10703
General	1910.316(a)	10703
Use	1910.316(b)	10703
Prohibited uses	1910.316(c)	10704
Splices	1910.316(d)	10704
Pull at joints and terminals	1910.316(e)	10704
Transformers	1910.318	10704
General	1910.318(a)	10704
Specific provisions applicable to different types of transformers	1910.318(b)	10704
Provisions for transformer vaults	1910.318(c)	10704
Appliances	1910.320	10705
General	1910.320(a)	10705
Installation of appliances	1910.320(b)	10705
Control and protection of appliances	1910.320(c)	10705
Hazardous locations, general	1910.322	10706
Application	1910.322(a)	10706

Heading	Subpart Section	Page
Special precaution	1910.322(b)	10706
Class I locations	1910.322(c)	10706
Class II locations	1910.322(d)	10706
Class III locations	1910.322(e)	10706
Class I installations—hazardous locations	1910.324	10707
General	1910.324(a)	10707
Transformers and capacitors	1910.324(b)	10707
Meters, instruments and relays	1910.324(c)	10707
Wiring methods	1910.324(d)	10707
Ceiling and drainage	1910.324(e)	10707
Switches, circuit breakers, motor controllers, and fuses	1910.324(f)	10708
Control transformers and resistors	1910.324(g)	10708
Motors and generators	1910.324(h)	10708
Lighting fixtures	1910.324(i)	10709
Utilization equipment, fixed and portable	1910.324(j)	10709
Flexible cords	1910.324(k)	10709
Receptacles and attachment plugs, Class I, Divisions 1 and 2	1910.324(l)	10709
Conductor insulation Class I, Divisions 1 and 2	1910.324(m)	10709
Signal, alarm, remote-control and communication systems	1910.324(n)	10709
Live parts, Class I, Divisions 1 and 2	1910.324(o)	10709
Class II installations—hazardous locations	1910.326	10710
General	1910.326(a)	10710
Transformers and capacitors	1910.326(b)	10710
Surge protection, Class II, Divisions 1 and 2	1910.326(c)	10710
Wiring methods	1910.326(d)	10710
Sealing, Class II, Divisions 1 and 2	1910.326(e)	10711
Switches, circuit breakers, motor controllers, and fuses	1910.326(f)	10711
Control transformers and resistors	1910.326(g)	10711
Motors and generators	1910.326(h)	10711
Ventilating piping	1910.326(i)	10711
Utilization equipment	1910.326(j)	10711
Lighting fixtures	1910.326(k)	10711
Class II, Division 1	1910.326(l)	10712

Heading	Subpart Section	Page
Receptacle and attachment plugs	1910.326(m)	10712
Signal, alarm, remote-control, and local loud-speaker, intercommunication systems	1910.326(n)	10712
Live parts, Class II, Divisions 1 and 2	1910.326(o)	10712
Class III installations—hazardous locations	1910.328	10713
General	1910.328(a)	10713
Transformers and capacitors, Class III, Divisions 1 and 2	1910.328(b)	10713
Wiring methods	1910.328(c)	10713
Switches, circuit breakers, motor controllers and fuses, Class III, Divisions 1 and 2	1910.328(d)	10713
Control transformers and resistors, Class III, Divisions 1 and 2	1910.328(e)	10713
Motors and generators	1910.328(f)	10713
Ventilating piping, Class III, Divisions 1 and 2	1910.328(g)	10713
Utilization equipment, fixed and portable, Class III, Divisions 1 and 2	1910.328(h)	10713
Lighting fixtures, Class III, Divisions 1 and 2	1910.328(i)	10714
Flexible cords	1910.328(j)	10714
Receptacles and attachment plugs, Class III, Divisions 1 and 2	1910.328(k)	10714
Signal, alarm, remote-control, and local loud-speaker intercommunication systems, Class III, Divisions 1 and 2	1910.328(l)	10714
Electric cranes and hoists and similar equipment, Class III, Divisions 1 and 2	1910.328(m)	10714
Electric trucks	1910.328(n)	10714
Storage-battery charging equipment	1910.328(o)	10714
Live parts	1910.328(p)	10714
Grounding	1910.328(q)	10714
Additional delay in effective date	1910.329	10714
Sources of standards	1910.330	10714
Standards organizations	1910.331	10714

CHAPTER TEN
Target Health Hazards Program

ASBESTOS

WHAT IS IT?

Asbestos is a general term used to describe several fibrous mineral silicates that occur as white, grayish or greenish masses, either compact or of long silky fibers. The most important types are Chrysolite®, a simple magnesium silicate; amosite, a complex magnesium iron silicate; and crocidolite, a complex sodium iron silicate. Some 95% of world asbestos production comes from Chrysolite.

WHERE IS IT?

The heat-resistant properties of asbestos have led to many uses, for example, protection against fire, insulation, brake and clutch linings, building materials, filters and in plastics. The raw material and end-products are found nearly everywhere.

WHAT IS THE HAZARD?

More than 200,000 employees face risks from asbestos. The principal danger is caused by inhaling asbestos fibers. Recent studies have shown the presence of asbestos fibers in the lungs of persons having no industrial exposure, probably due to their presence in the atmosphere near construction sites. Prolonged inhalation of asbestos fibers between 5 and 50 microns long (one-millionth of a meter) can produce a lung disease known as asbestosis. The lungs cannot eliminate asbestos fibers, so the fibers coat the tissues. If this process continues over a period of 10 to 20 years, with significant quantities of asbestos present, the tissue reaction progresses until a generalized, diffuse fibrosis occurs, causing severe respiratory disability. There have been reports of increased incidence of lung cancer in persons with asbestosis.

HOW IS IT CONTROLLED?

Enclosure and local exhaust ventilation applied to equipment and/or operations are the principal means of preventing employee exposure to dangerously high concentrations. The U. S. Bureau of Mines states that approved dust respirators may be worn by employees as protection for some operations.

WHAT ARE APPROVED LEVELS?

OSHA's permissible level is five fibers per milliliter greater than five microns in length for an eight-hour, time-weighted, average airborne concentration. This may be increased to 10 such fibers per milliliter for no more than 15 minutes per hour, up to five hours per eight-hour day. Imminent danger situations are generally not applicable. Any exposure greater than permissible levels for unprotected or improperly protected workers is considered a serious violation.

HOW IS IT MEASURED?

Asbestos dust will be collected with a personal sampling pump. Fibers will be counted microscopically at 400-450 magnification using phase contrast illumination.

CARBON MONOXIDE

WHAT IS IT?

Carbon monoxide is a colorless, almost odorless, tasteless, non-irritating gas. It is produced by the incomplete combustion of fuels, such as wood, gas, coal, oil and gasoline. Miners refer to it as "white damp." It also is commonly called "coal gas."

WHERE IS IT?

Carbon monoxide is found anywhere carbonaceous fuels are used. It can be found in everyday activities as well as in industrial situations, near incorrectly vented gas heaters, near gasoline-powered machinery and around automobiles run indoors, blast furnaces, and catalytic reduction operations.

WHAT IS THE HAZARD?

Because it is found so widely, nearly everyone, at one time or another, is exposed to carbon monoxide. The danger lies in inhalation of the gas. Carbon monoxide enters the blood stream through the lungs in the same manner as oxygen, but it displaces oxygen in the blood because it has approximately 210 times greater affinity for hemoglobin in the blood than oxygen. As a result, the blood cannot transport oxygen, causing suffocation even in an atmosphere containing ample oxygen. Over a period of time, as little as 700 parts per million of carbon monoxide will saturate 50% of the blood hemoglobin, while a few breaths at concentrations of 10,000 parts per million may cause 60 to 80% saturation and death. Fortunately, regardless of how high the percentage of saturation of the blood has been, practically all carbon monoxide is eliminated by the body within 8 to 10 hours after exposure ends.

HOW IS IT CONTROLLED?

In general, adequate ventilation will prevent carbon monoxide poisoning. Enclosure, general or dilution ventilation and local exhaust ventilation, as well as isolation of hazardous operations are the principal means of preventing employee exposure to dangerous concentrations. Automated detection alarm systems also are useful control measures.

WHAT ARE APPROVED LEVELS?

OSHA's permissible level is 50 parts per million parts of air for an eight-hour, time-weighted, average airborne concentration. Concentrations above 500 parts per million for any length of time are considered to be imminent danger situations. Concentrations above 150 parts per million for a total of one hour or more are considered to be serious violations. Concentrations above 50 parts per million, but below 150 parts per million, for an eight-hour, time-weighted average are considered regular or nonserious violations.

HOW IS IT MEASURED?

The primary method of measurement is a carbon monoxide portable, Hopcalite-reagent type meter. The secondary method is certified detector tubes.

COTTON DUST

WHERE IS IT?

Cotton dust occurs throughout the textile industry and affects cotton, flax and soft hemp workers. The greatest exposure to cotton dust generally occurs during the carding of cotton, an operation that combs out cotton fibers and sets them parallel by drawing them between toothed cylinders. However, machinery throughout the cotton mill must be cleaned frequently, and much exposure to cotton dust occurs then.

WHAT IS THE HAZARD?

More than 800,000 employees in cotton processing operations of all types are exposed to cotton dust. While the exact effects of cotton dust are not known, prolonged exposure to heavy air concentrations can cause a disabling lung disease known as byssinosis. Materials adhering to cotton fibers, such as unwanted leaf, stem and boll trash, when inhaled, elicit an allergic reaction or response. Byssinosis generally advances through three recognizable stages. The first is commonly called "Monday fever" and begins only after exposure of more than 10 years. Workers develop an irritating cough, tightness of the chest and breathlessness. The attacks usually come on Monday, or after a period of absence from the plant and last only a day or so; hence the name "Monday fever." In the second stage, these symptoms extend over more days in the week and finally become permanent. If exposure to cotton dust is ended at this stage, recovery can occur. In the third stage, the tightness of the chest and labored breathing become so distressing the worker must leave the industry. The disease can progress then to chronic bronchitis and emphysema.

HOW IS IT CONTROLLED?

Enclosure of machinery in all operations and local exhaust ventilation are the principal means of preventing employee exposure. Pretreatment of the cotton in early processing stages may also be helpful.

WHAT ARE APPROVED LEVELS?

OSHA's permissible level is one milligram per cubic meter of air for an eight-hour, time-weighted, average airborne concentration. Imminent danger situations are generally not applicable. Concentrations of three milligrams per cubic meter are considered serious violations; concentrations between one and three milligrams per cubic meter for an eight-hour, time-weighted average are considered regular or nonserious violations.

HOW IS IT MEASURED?

A personal sampler is attached to a workman's clothing with the uncovered filter facing down. Usually a six hour sample is obtained.

LEAD

WHAT IS IT?

Lead is a soft, dense, malleable, heavy metal with a low melting point. It is gray in color and, because of its characteristics, has a wide application of uses.

WHERE IS IT?

Lead is used to make printing type and plumbing, to shield telephone and power cables and to protect against radiation hazards. As an alloy, lead mixed with steel makes excellent bearings. As solder, lead alloy is vital in the electrical and electronic industries. Lead compounds are used to make paints and ammunition, and lead oxides are used in car batteries.

WHAT IS THE HAZARD?

More than 1.6 million employees in nearly every industrial manufacturing process and in many service industries are exposed to the poisoning effect of lead. Lead can enter the body by inhalation, absorption through the skin or by ingestion. In sufficient quantities, the result is the same—severe gastrointestinal, blood and central nervous system disorders. Inhalation of lead dust or fumes is the most frequent means of entry and results in most of the industrial health problems involving this metal. Because lead is a cumulative poison, a part of a small daily dose is not eliminated but is stored in the body. Eventually, a point is reached where symptoms and disability, even death, occur.

HOW IS IT CONTROLLED?

Since the primary route of industrial lead poisoning is inhalation, enclosure and local exhaust ventilation are the principal means of control. Daily wet or vacuum cleaning of all lead dust, personal hygiene, prohibition of food and beverages in the work area and use of U. S. Bureau of Mines-approved respirators are other means that can help control exposure.

WHAT ARE APPROVED LEVELS?

OSHA's permissible level is 0.2 milligrams per cubic meter of air for an eight-hour, time-weighted, average airborne concentration. Imminent danger situations are not generally applicable. Any exposure above 0.6 milligrams per cubic meter of air for an eight-hour, time-weighted average is considered a serious violation. Levels above 0.2 and less than 0.6 are considered regular or non-serious violations.

HOW IS IT MEASURED?

The sample is collected for a minimum of 60 minutes by a personal sampling pump on a millipore filter. The filter is subsequently dissolved in nitric acid by taking the lead into solution. The solution is then introduced into an atomic absorption unit for analysis.

SILICA

WHAT IS IT?

Silica is an oxide of silicon and is the characteristic ingredient of a great variety of minerals, including quartz, cristobalite, tridymite, amethyst and sand. It is widely used in cores for metal casting operations, ceramics, decorative materials, insulation, building materials and glass.

WHERE IS IT?

Silica is found throughout industry in various manufacturing processes, such as iron and mineral processing plants, mining, abrasive soap production, glass manufacturing, potteries, foundries, abrasives use and manufacturing, in stone cutting and finishing industries.

WHAT IS THE HAZARD?

More than 1.1 million employees are exposed to the hazard created by inhalation of silica dust. Inhalation of these particles can produce rapidly-developing (acute) or chronic silicosis, a disabling lung disease. Typically, the lung tissue reacts to the presence of silica particles by forming fibrous matter around the particle. Evidence suggests that particles below one micron (one-millionth of a meter) in size may be the most dangerous, since they may penetrate deeper into the lungs in high concentrations. In rapidly developing silicosis, symptoms appear eight to 18 months after the first exposure. Chronic silicosis, the type usually encountered in industry, generally is produced only after years of silica inhalation. The symptoms for either type are the same—slowly increasing difficulty in breathing under exertion. In some workers, large masses of dense fibrous tissue develop in the upper portion of the lungs, leading to severe respiratory crippling. Tuberculosis is a frequent complication of silicosis.

HOW IS IT CONTROLLED?

The best method is substitution of another material; for example, use of steel shot or alumina grit in place of sandblasting operations. Other methods include enclosure of operations, local exhaust ventilation, segregation of operations, wetting, and ordinary good-housekeeping practices.

WHAT ARE APPROVED LEVELS?

For respirable quartz particles, OSHA's permissible level is 10 milligrams per cubic meter of air, divided by the percent of free silica plus 2, as shown in this formula: $\dfrac{10 \text{mg/m}^3}{\% \text{SiO}_2 + 2}$.

For respirable cristobalite and tridymite, the limits are half the value for quartz. Imminent danger situations generally are not applicable. Any eight-hour, time-weighted average greater than four times the limit is considered a serious violation. Exposure to concentrations between the limit and four times the limit are considered a regular or nonserious violation.

HOW IS IT MEASURED?

Personal sampling pumps with filters are used to collect dust samples for worker exposure. The collected dust then is weighed for a given volume of air. Similar procedures are used to determine the percentage of silica in the air.

CHAPTER ELEVEN
NIOSH

Section 22 of the Act creates a National Institute of Occupational Safety and Health (NIOSH) in the Department of Health, Education and Welfare (HEW) to carry out Congressional policy, as set forth in Section 2 of the Act, to develop and establish recommended occupational safety and health standards, and to perform the research, training and other functions of the Secretary of Health, Education and Welfare delineated in Sections 20 and 21 of the Act.

The establishment of the Institute is intended to "elevate the status of occupational health and safety research to place it on equal footing with the research conducted by HEW into other matters of vital social concern."

In addressing an AFL-CIO conference on the Occupational Safety and Health Act, Dr. Marcus Key, first Director of the Bureau of Occupational Safety and Health, in HEW, outlined the organizational structure, functions and responsibilities of the new structure.

FUNCTIONAL ORGANIZATION

NIOSH is organized functionally, rather than along disciplinary lines. There are operating divisions for each major activity. For health and safety research there is the Division of Laboratories and Criteria; for industry-wide studies the Division of Field Studies and Clinical Investigation; for hazard evaluations the Division of Technical Services; and for short-term training the Division of Training.

These four divisions operate out of Cincinnati, Salt Lake City, and the regional offices. Their directors report to an Associate Director in Cincinnati who reports to the Director in Washington. The Appalachian Laboratory for Occupational Respiratory Diseases in Morgantown has the sole responsibility for implementing the Coal Mine Health and Safety Act.

A division in Washington promotes occupational health programs at the Federal and state levels as well as in industry and agriculture. This division also has the very important responsibility of correlating the practice of occupational medicine in industry with the total delivery of health services. Other offices in Washington that interface with the Department of Labor include the Office of Health Surveillance and Biometrics, the Office of Manpower Development, and the Office of Research and Standards Development.

NIOSH technical services operations are decentralized and operate on the regional basis, similar to OSHA function and structure.

Plans for the end of 1972 were to staff regional offices with two professionals—a program director and an assistant and a secretary. Original staffing was estimated on the basis of the expected number of requests for hazards evaluations and in all probability will be increased.

Monitoring and surveillance activities are also part of the NIOSH regional office function. The office employs industrial hygienist surveyors who conduct preliminary surveys to gather information on the industrial hygiene problems in the United States. Regional offices also provide assistance on State plans, contract and grant monitoring, training and recruitment.

HEW RESPONSIBILITIES

The major responsibilities of HEW, according to Dr. Key, are these: health and safety research, industry-wide studies, hazard evaluations, toxicity determinations, annual listing of toxic substances, record-keeping, and manpower development and training.

Of the many different types of research which are authorized for HEW, the most important is that which produces criteria on which standards for toxic materials and harmful physical agents can be based. The criteria which HEW will transmit to the Department of Labor will contain not only all pertinent dose-effect studies with literature citations, but also specific limit recommendations based on health effects.

Feasibility studies are performed to produce a manual of good practices for each substance, with the Department of Labor adding items relating to the national interest and economic considerations for publishing proposed standards, and for public hearings if necessary.

"HEW," Dr. Key said, "has had considerable experience in this type of health research, but we have a big task ahead of us in tooling up for safety research. In this area we plan to emphasize motivational and behavioral factors in safety, ergonomics and work physiology."

NIOSH is charged, within two years of its inception and annually thereafter, with conducting and publishing industry-wide studies on the effect of chronic and low-level industrial exposures which may have potential for disease or physical impairment.

The asbestos, beryllium, cotton dust and noise studies first undertaken by NIOSH are examples of the prescribed industry-wide studies. These environmental-medical studies will increase and many different types of studies and records will be added as NIOSH looks for long-term effects and for causes of shortened life span.

TOXICITY RULINGS

NIOSH is required to make hazard evaluations and toxicity determinations when a request in writing is made by any employer or group of employees. Estimates at the outset of the program were that there would be 5,000 such requests a year.

The requests will be received through regional offices which have access to a computerized technical information system. In this system are stored trade name, composition, toxicity data, sampling and analytical procedures, and suggested limits, if any.

If the request concerns a substance for which there is already a standard, it is referred to the Department of Labor. Otherwise a field visit by an industrial hygienist from NIOSH is made. Anticipation is that final determination can be made on the first visit for about half of the requests, field sampling on a return visit for some 25%, and the remainder, in the area of almost complete unknowns, will require acute, subacute and chronic toxicity studies. Medical studies may also be indicated.

If there are more requests than can be handled, priorities will be established based on the severity of risk and numbers of people exposed.

The annual compilation of toxic substances will identify each substance by generic name, will give the known toxic concentration which has most application to man under conditions of occupational exposure, and will contain literature citations. For many substances there will be no human data and it will be necessary to supply animal data with the closest approximation to the industrial route of exposure given.

The definition of toxicity is broad, including carcinogens (cancer-producing), mutagens (change-producing) and teratogens (monster-producing). Estimates are that some 12,000 items will comprise the first list, compared with the 400-500 substances for which TLV's (threshold limit values) or standards are available. This is considered to be NIOSH's "shopping list" for standards setting. Later lists will be longer and contain much more information.

Record keeping will be required (of employers) by the Department of Labor, and in some cases, HEW. HEW requests, however, will be made through the Department of Labor so as not to confuse or overwhelm industry. In the development of the record keeping and reporting system NIOSH has worked with the Department of Labor and ANSI (American National Standards Institute). Occupational illnesses will be reported for the first time, and will be broken down into several major categories.

EDUCATIONAL PROGRAMS

One of NIOSH's principal responsibilities is the development of educational programs to provide an adequate supply of qualified personnel to carry out the purposes of the Act. Such trainees would be utilized by the Department of Labor, the Department of HEW, other Federal agencies, state health and labor departments, business, industry and political subdivisions.

Implementation of the manpower responsibility includes many different programs. Through short courses, NIOSH says, and the associated extern programs in industry and state occupational safety and health agencies, it will train high school graduates to be technicians, junior college graduates to be para-professionals, and college graduates to be occupational safety and health professionals.

NIOSH seeks to increase the number of occupational safety and health professionals, para-professionals and technicians through
1. two to four year college program support
2. nursing school support for a last year option in occupational health nursing
3. masters level graduate training
4. research training grants.

Dr. Key stated, and experience to date tends to bear him out, that the health aspects of the Act will overshadow the safety aspects, and that HEW responsibilities are much more difficult than DOL responsibilities.

Enforcement, he said, is relatively easy when the limit is given and the sampling method and analytical procedures are specified. The greater challenge and responsibility lies in trail blazing or ascertaining if there is a health hazard, programming criteria or dose-effect type research to set a standard, and determining the limit which offers the greatest health protection.

ADMINISTRATIVE RESPONSIBILITY

According to its organizational statement NIOSH assumes responsibility for:
1. administering health and safety research, including studies of psychological factors pertaining to employment health and safety,
2. developing innovative methods and approaches for dealing with job-related safety and health problems,
3. providing medical criteria to provide, "insofar as practicable, that no employee will suffer diminished life expectancy as a result of his work experience," to develop methods of recognizing latent diseases and to isolate causal relationships between diseases and working conditions,
4. contributing to the development of training programs to increase the numbers and competence of safety and health personnel,
5. developing reporting procedures to aid in describing the nature of the nation's safety and health problems, and,
6. consulting with the U. S. Department of Labor, other federal, state and local government agencies, labor and management groups, and others interested in the promotion of occupational safety and health.

Administrative responsibility for these activities has been divided among eight main offices and several operating divisions within the National Institute. The principal divisions and duties are described below.

OFFICE OF THE DIRECTOR

This office (1) plans, directs, coordinates, and evaluates the operations of the institute, (2) maintains liaison with and provides advice and assistance to the U.S. Department of Labor, the U.S. Department of the Interior, other Federal agencies, state and local government agencies, international health organizations, and outside groups, (3) provides coordination with the Federal Health Programs Service's occupational health activities for Federal employees, and (4) provides policy guidance and coordination to occupational safety and health activities in the Regional Offices.

OFFICE OF PUBLIC INFORMATION

This office (1) assists and advises the Institute Director and the Divisions on public information policies and activities, (2) provides information materials for response to public inquiries, (3) coordinates printing, publication, and clearance procedures for the Institute, and (4) assists in developing displays, exhibits, and illustrations.

OFFICE OF EXTRAMURAL ACTIVITIES

This office (1) advises the Institute Director on matters relating to the development and progress of Institute-supported external research, (2) in cooperation with the offices and operating divisions of the Institute, stimulates research, training, and demonstration grants in relevant priority areas, and (3) administers the management aspects of the Institute's grants programs by receiving, reviewing, analyzing, and evaluating all grant applications.

OFFICE OF ADMINISTRATIVE MANAGEMENT

This office (1) provides management information, advice, and guidance to the Institute Director, (2) coordinates all management activities in the conduct of finance, personnel, and procurement functions, (3) relates administrative management activities to programs, and (4) develops necessary policies, procedures, and operations, and provides such special reports and studies as may be required in the management area.

OFFICE OF PLANNING AND RESOURCE MANAGEMENT

This office (1) plans and coordinates the strategy and philosophy of operation of the Institute regarding mission and objectives, (2) conducts or participates in special studies for program planning and evaluation, (3) conducts the necessary control functions to assure operational compliance toward program objectives within the Institute, and (4) provides management systems consultation and analyses.

OFFICE OF RESEARCH AND STANDARDS DEVELOPMENT

This office (1) reviews existing scientific criteria for health and safety standards and assesses through priority systems the needs for additional research program areas for criteria development, and (2) coordinates and maintains an overview of research activities in the operating divisions of the Institute with the ultimate aim toward finalization of criteria and standards.

OFFICE OF MANPOWER DEVELOPMENT

This office (1) provides policy guidance and evaluates the Institute's manpower development and training activities, (2) advises the Institute Director on national health manpower needs related to occupational safety and health, and relates to other Federal agencies regarding occupational safety and health manpower needs, and (3) conducts equal employment opportunity activities of the Institute as part of the total HSMHA-EEO program.

OFFICE OF HEALTH SURVEILLANCE AND BIOMETRICS

This office (1) operates as the principal statistical and data research unit in the Institute, (2) monitors new as well as existing occupational hazards, and maintains surveillance on the incidence of occupational illnesses and diseases, (3) in coordination with the U. S. Department of Labor, establishes a priority list for the conduct of research and the development of standards, (4) develops and conducts record studies of work population groups to determine the national trends and problem areas related to job health and safety, and provides health policy guidance in epidemiology, and (5) coordinates the Institute's electronic data processing requirements, to ensure that adequate computer facilities and services are available.

DIVISION OF LABORATORIES AND CRITERIA DEVELOPMENT

This office (1) develops criteria for standards for the control of chemical, biological, and physical hazards to the health and safety of the working population, and initiates standard methodology and instrumentation for the detection, evaluation, and control of such hazards, (2) evaluates the toxicity, health, and safety hazards of industrial substances, processes, and other agents, as well as current research requirements and regulations, (3) conducts methodology studies for evaluating the varying capacity of workers to withstand physical and psychological responses, (4) provides for equipment development, analytical service, and calibration needs of other operating divisions within the Institute, and maintains an analytical and calibrations service for the U. S. Department of Labor, and (5) evaluates and certifies the performance of safety and health equipment.

DIVISION OF FIELD STUDIES AND CLINICAL INVESTIGATORS

This division (1) conducts nation-wide studies, surveys and comprehensive analyses to determine the health status of the working population, including the incidence and prevalence of disease and injury, and (2) initiates studies to determine chronic and long-term effects of work-related exposures to toxic and hazardous substances.

DIVISION OF TECHNICAL SERVICES

This division (1) provides demonstrations, technical assistance, and consultation to public and private agencies responsible for the control of occupational diseases and accidental work injuries, (2) through the Regional Offices and its central staff serves as the focal point for the review of state plans and grants with the U. S. Department of Labor and makes the initial responses to requests for hazards evaluations, (3) in cooperation with the Office of Extramural Activities, stimulates, programs, and monitors demonstration grants for new and innovative methods of recognizing, evaluating, and controlling occupational hazards, (4) prepares manuals of good practice for safe work procedures, and (5) operates the technical information inquiry service of the Institute.

DIVISION OF OCCUPATIONAL HEALTH PROGRAMS

This division (1) promotes occupational health programs at the state and local governmental levels as well as in industry and agriculture, (2) provides technical guidance in the development of occupational health programs, and (3) correlates the practice of occupational medicine in industry with the total delivery of health services.

DIVISION OF TRAINING

This division (1) develops and plans short-term training activities for Federal, state, and local governments, industry, and other appropriate organizations in the field of occupational safety and health, and (2) conducts such short-term training.

APPALACHIAN LABORATORY FOR OCCUPATIONAL RESPIRATORY DISEASES

The laboratory (1) conducts studies of the incidences and prevalence of occupational respiratory diseases in specific work groups with particular emphasis on coal workers' pneumoconiosis, and (2) provides medical and engineering research and service to fulfill the Institute's responsibilities under the Federal Coal Mine Health and Safety Act of 1969.

1972 PRIORITY LIST FOR CRITERIA FOR TOXIC SUBSTANCES AND PHYSICAL AGENTS

The following list was designed to help researchers identify areas where criteria information is needed. Five weighted indices were

assigned to each substance under consideration. They were rated on a scale of 9 to 1 and included the following:
1. Population Index (PI): a relative evaluation of worker exposure
2. Relative Toxicity Index (RT): professional opinion on relative toxicity
3. Incidence Index (II): occupational disease records, physicians' first reports, workmen's compensation records, and other documented incidents of disease
4. Quantity Index (QI): amount produced or used each year
5. Trend Index (TI): estimates on increased or decreased usage.
The overall rating for each substance was computed using the formula: 2PI + 2RT + II + QI + TI divided by 5.

NATIONAL INSTITUTE FOR OCCUPATIONAL SAFETY AND HEALTH PRIORITY LIST FOR CRITERIA FOR TOXIC SUBSTANCES AND PHYSICAL AGENTS, 1972

Criteria have been developed for asbestos and coal dust. Substances for which criteria are still being developed are arsenic, benzene, beryllium, cadmium and compounds, carbon monoxide, chromic acid mist, cotton dust, fibrous glass, heat stress, lead, mercury, noise, parathion, silica, trichloroethylene, and ultraviolet.

PRIORITIES

1.
Bis (chloromethyl) ether
Coal tar pitch volatiles
2-Naphthylamine
Toluene diisocyanate
Radioactive products of uranium mining (gaseous and particulate)

2.
Benzidine and its salts
Carbon Tetrachloride
Ozone
Sulfur dioxide
Tin and compounds

3.
Chromium compounds
Dichlorobenzidine
Oxides of nitrogen
Sodium hydroxide
Sulfuric acid

4.
Carbaryl
Chloroform
4-Dimethylaminoazobenzene
Nitric acid
Toluene

5.
Ammonia
Beta-propiolactone
Epoxy resins
Methylene chloride
4-nitrodiphenyl

6.
Asphalt fumes
Ethylene dichloride
Fluoride and HF
Polychlorinated biphenyls
Tetrachloroethylene

7.
2-Acetylaminofluorene

Chlorobenzene
Methylene bisphenyl isocyanate (MDI)
Phosgene
Trichloroethane
8.
Acetone
4-Aminodiphenyl
Dieldrin
Malathion
N-nitrosodimethylamine
9.
Aniline
Copper and compounds
Cyanides
Styrene
Zinc and compounds
10.
Chlorine
Formaldehyde
Manganese and compounds
Phenol
Platinum and compounds
11.
Acrolein
Aluminum and compounds
Carbon disulfide
Methyl ethyl ketone
Vinyl chloride
12.
Creosote
Methyl chloride
Nickel and compounds
Phosphorus and compounds
Tetrachloroethane
13.
Acrylonitrile
2,4-Dinitrophenol
Magnesium and compounds

Methyl alcohol
Paraffin
14.
Ammonium nitrate
Cold stress
Dioxane
Flourine
Microwaves
15.
Hydrogen chloride
Ethyl benzene
Nitroglycerin
Vibration
Xylene
16.
Methyl butyl ketone
Mineral spirits
Oil mists
Selenium and compounds
Turpentine
17.
Arsine
Gasoline
Kerosene
Iron and compounds
Petroleum naphtha
18.
Barotrauma
Cresol
Paraquat
Portland cement
Talc
19.
Carbon black
Coherent energy (laser radiation)
Ethylene oxide
Impact noise
Proteolytic enzymes

NATIONAL SURVEILLANCE NETWORK
NATIONAL OCCUPATIONAL HAZARD SURVEY FORM

This is included in the text to give employers and employees some specific idea of the scope and function of NIOSH and the importance attributed to the "health" side of the Occupational Safety and Health Act of 1970.

1. Facility name
2. Address
3. City, state, zip code
4. Legal Owner(s)
5. Person(s) interviewed: name, title, part
6. Area code, telephone
7. Date survey started or visit made
8. Facility identifier
   If mailing address or person to contact concerning information about the survey is different from that indicated above, list the correct mailing and contact below.
9. Facility name
10. Address
11. City, state, zip code
12. Attention: title
13. Area code, telephone

Facility Health and Safety Services
1. Revision code
2. Date survey started (month/day/year)
3. State code
4. Facility identifier code
5. Area of business
   a. What is your major activity?
   b. What are your chief products, services, lines of trade, etc.?
   c. SIC code if known.
6. Approximately how many years has this facility been involved in this activity?
7. How many people are on your facility payroll for all shifts at the present time?
8. Of this number, how many are normally in the work areas as opposed to the administrative or other areas?
9. How many shifts do you have?
10. Has this facility received industrial hygiene services during the past year?

Yes, from an industrial hygienist (1)
Yes, from a safety engineer (2)
  (Skip to Question 13)
Yes, from other. Specify: (3)
  (Skip to Question 13)
No   (Skip to Question 13) (4)

11. Is the industrial hygienist based in this facility?
Yes (1)

No, consulting basis (3)
No, other. Specify: (4)
Not applicable (5)

12. Estimate the average number of industrial hygienist hours that are devoted to your facility per month.

13. Has your facility received safety engineer services during the past year?
Yes, from a safety engineer based in this facility (1)
Yes, from a safety engineer based elsewhere (2)
Yes, on a consulting basis (3)
Yes, other. Specify: (4)
No (5)

14. Estimate the average number of safety engineer hours that are devoted to your facility per month.

15. Is there a formally established health unit at this facility?
Yes, physician in charge (1)
Yes, Registered Nurse in charge (2)
Yes, Licensed Practical Nurse in charge (3)
Yes, other in charge. Specify: (4)
No (5)

16. Do you employ or have an arrangement with a physician or clinic to give your employees medical care?
Yes, employed full time (1)
Yes, employed part time (2)
Yes, on call   (Skip to Question 18) (3)
Yes, at clinic   (Skip to Question 18) (4)
Other. Specify: (Skip to Question 18) (5)
No arrangements made   (Skip to Question 18) (6)

17. Estimate the average number of physician hours that are devoted to your facility per week.

18. Do you have one or more nurses at this facility to provide care for employees?
Yes (1)
No (Question 19—not applicable) (2)

19. How many Registered Nurses and Licensed Practical Nurses are employed at this facility?
20. Estimate the average number of nursing hours that are devoted to your facility per week.
21. Do you have an employee at this facility with formal first-aid training, other than doctors or nurses, who has been designated to provide emergency treatment?
    Yes (1)
    No (2)
22. Do you record health information about a new employee on some regular form?
    Yes, all employees (1)
    Yes, executive and/or managerial only (2)
    Yes, other employees (3)
    No (4)
23. Before new employees are hired or placed, are they required to take a medical examination?
    Yes, all employees (1)
    Yes, executive and/or managerial only (2)
    Yes, other employees (3)
    No (4)
24. Do you require medical examinations of your employees who return to work after a sickness, or whose employment is terminated?
    Return to work only (1)
    Exit examination at time of termination only (2)
    Both (3)
    Neither (4)
25. Do you provide periodic medical examinations or tests of any type for employees?
    Yes, all employees (1)
    Yes, executive and/or managerial only (2)
    Yes, other employees (3)
    No (4)
26. Do you provide periodic ophthalmologic examinations for employees?
    Yes, all employees (1)
    Yes, executive and/or managerial only (2)
    Yes, other employees (3)
    No (4)
27. Do you provide periodic audiometric examinations for employees?
    Yes, all employees (1)

       Yes, executive and/or managerial only (2)
       Yes, other employees (3)
       No (4)
28. Do you provide periodic blood tests for employees?
       Yes, all employees (1)
       Yes, executive and/or managerial only (2)
       Yes, other employees (3)
       No (4)
29. Do you provide periodic urine tests for employees?
       Yes, all employees (1)
       Yes, executive and/or managerial only (2)
       Yes, other employees (3)
       No (4)
30. Do you provide periodic pulmonary function tests for employees?
       Yes, all employees (1)
       Yes, executive and/or managerial only (2)
       Yes, other employees (3)
       No (4)
31. Do you provide periodic chest X-rays for employees?
       Yes, all employees (1)
       Yes, executive and/or managerial only (2)
       Yes, other employees (3)
       No (4)
32. Do you have a regularly scheduled program to give employees flu, cold, or other immunizations?
       Yes (1)
       No (2)
33. Do you keep employee absenteeism records?
       Yes, showing specific nature of sickness when absent (1)
       Yes, showing only the type of absence (2)
       Yes, without showing nature of absence (3)
       No (4)
34. What is your rate of unscheduled absenteeism (days per year per employee)?
35. Is there a formally established safety committee at your facility?
       Yes, investigative (1)
       Yes, policy setting (2)
       Both (3)
       Yes, other. Specify: (e.g. advisory only) (4)
       No (5)

36. Do you have areas where personal protective devices are required or recommended?
    Yes, required (1)
    Yes, recommended (2)
    Yes, both (3)
    No (4)
37. Who provides personal protective devices?
    Employee (1)
    Employer (2)
    Both (3)
    Other. Specify: (4)
    Not applicable (5)
38. Who has been designated to see to it that personal protective devices are serviced and maintained?
    Individual employees (1)
    Employer representative (2)
    Both (3)
    Other. Specify: (4)
    No one (5)
    Not applicable (6)
39. How do you carry your workmen's compensation insurance?
    Private insurance company (1)
    Self-insured (2)
    State insurance fund (3)
    Other. Specify: (4)
    None (5)
40. Are any unions operating in this facility?
    Yes (1)
    No (2)
41. Is this facility a member of a national association or institute representing its industry or trade?
    Yes (1)
    No (2)
42. Do you have a program under which you regularly monitor the presence of fumes, gases, mists, vapors, dusts, noise, vibration, radiation or other similar conditions?
    Yes (1)
    No (2)
43. Do you use a private sewage treatment plant or a septic tank to dispose of this facility's nonprocess sewage?
    Yes, private sewage treatment plant (1)
    Yes, septic tank (2)

Yes, both (3)
No (4)

44. Do you have any drinking water other than that provided through a public water supply?
    Yes. Specify: (1)
    No (2)

45. Is the private water supply routinely tested for bacteriological quality?
    Yes (1)
    No (2)
    Not applicable (3)

46. Do you have any piped liquids other than drinking water?
    Yes. Specify: (1)
    No (2)

47. Is there a separate, identified eating area for your work area employees?
    Yes (1)
    No (2)

48. Are hand washing facilities provided within 200 feet of the eating area?
    Yes (1)
    No (2)
    Not applicable (3)

49. May I see the Summary of Occupational Injuries and Illnesses (OSHA Form 102)?
    Occupational injuries
    Occupational illnesses
    Occupational skin diseases or disorders
    Dust diseases of the lungs (pneumoconioses)
    Respiratory conditions due to toxic materials
    Poisoning (systemic effects of toxic materials)
    Disorder due to physical agents (other than toxic materials)
    Disorders due to repeated trauma
    All other occupational illnesses. Specify:

50. How many months are covered by the preceding figures?

CHAPTER TWELVE
Conclusion

The Act, of course, poses business, industry and political subdivision problems. Certainly it adds to the numbers of regulatory mandates which business claims makes the profit-making process more and more difficult. The truth of this is difficult to assess.

The Act's necessity, however, just like traffic laws or other regulations in our overall group and individual interest, seems unquestionable. It seems fairly self-evident that with the lack of safety and health laws of any description in some states, with fair laws in others, and even with good laws lacking ordinances for enforcement, or finding those ordinances not implemented—that something had to happen with sufficient national impact to put environmental occupational protection in proper perspective.

The Williams-Steiger Occupational Safety and Health Act of 1970 is a good beginning, but it gave some indication in its first year of implementation of being a political document, subject to the winds of change that such documents incur.

The Act, its language, intent and purpose is noble and majestic in tone, objective and ideal, and is, in fact, long overdue. What we do with it here in the U. S. is of concern to those who have invested their lifetime, their talent, their ambition and whatever dedication they could manage, to the proposition that the working man and woman has every right to expect that life on the job should be free (not as much as "feasible" or "practicable," but as much as possible) from injury, death or illness.

It remains to be seen whether a vested interest group, and labor is no different than employer in this area, pulls the teeth of what began as a healthy jawful of molars, grinders and incisors. Some *visits to the dentist* have already occurred.

The closest thing to a real representative for the average shift worker, hard-hat, blue collar or clock puncher—the individual who gets the job done—is not the union nor the employer. It is the Federal government when it enacts the kind of legislation represented by the Williams-Steiger Occupational Safety and Health Act of 1970.

The employee stands between a rock and a hard place when these two giants argue and negotiate over his welfare or lack of it, what is best for him and what isn't. Decisions that can make the difference between what is really a safe and healthful practice or place to work and what is not can find their way no farther than a review committee, a criteria package, a covert or overt power play, a politician seeking re-election with numerous "obligations" to repay, before they disappear into oblivion.

The Act makes this possibility more difficult, but there have been some alarming back-offs in the face of all that brave and wonderful wordage, and the tendency to use the *weasel words* is still prevalent.

The Act's biggest problem, however, in terms of securing its stated objectives, is not political. It is educational. It is securing and training personnel in numbers and capability to effectively manage this new responsibility. Involved as we have been in developing a training and education program for the University of Puget Sound, our observation is that the answer is not the Act itself but the people responsible for its leadership in the workplace. The supply of such trained personnel is woefully short and will be for some time.

The reason, of course, is the limbo in which something vaguely defined, known and shelved as "safety" has found itself for years. If "safety" was relegated to that position, such sophisticates as "Occupational Health" and "Industrial Hygiene" were even lesser known.

Seventy per cent of today's industry, companies who hire less than 100 employes, still know little about the Act and much less about NIOSH, or what industrial hygiene or occupational health has to do with them.

NIOSH, as we noted, faced with the responsibility of "education and manpower development," is adding training programs. HEW is supporting occupational safety and health curriculi with federal funding, as colleges and universities for the first time are developing and presenting training programs to meet a need which has been there for a long time.

In the press and rush to meet this need, there is a tendency to simply cram a curriculum together and start teaching it. HEW funding requirements, as a matter of fact, make it virtually impossible to *develop* a curriculum. If a college, university or institute is seeking federal funds to initiate an environmental occupational protection curriculum, in order to qualify for such funding the curriculum must be already completely determined—courses, subject matter, credit hours, etc. This has some inherent and built-in handicaps. First of all, from any logical viewpoint, a curriculum has to be *developed,* particularly a new one. It needs to be developed to meet the need, not the pragmatic specification of some one or some group who says that curriculum meets the need.

Today's occupational safety and health requirements, rules and regulations, posture, attitude, philosophy, objectives and goals, is a new ball game. Stereotyped, standard, traditional inflexibilities need their clanking machinery unlocked. New definitions, new wordage, new application, and new horizons are needed. Such words as "safety" and "accident" need to be changed, redefined, or eliminated. Occupational safety and occupational health need to become a partnership rather than two separate and competing disciplines.

The real, immediate and urgent training need is not two-year, four-year, or graduate level degree programs. It is short-course training for the man or woman faced with the new responsibility of coping with OSHA—1970. He needs competent, capable, instruction directly related to the Act and its requirements.

Such courses need to be developed in subject matter and time-span so they can be applied for later credit within a two-year or four-year or graduate degree program. Out of such development and experience an intelligent, practical and working curriculum can be developed.

While federal funding of such projects leaves much to be desired, the plain and simple fact is that any curriculum to be successful must be self-sustaining. Curricula already set in concrete tends to follow a lead set in "safety" in years preceding the passage of the Act, years singularly unsuccessful in any noteworthy achievement other than holding the line. Also today's "professional" is still sitting on his hands, waiting for someone else to take the lead, is still doing someone else's bidding, is still looking to management for approval before making any kind of move.

In terms of opportunity to do a better and more professional environmental occupation protection job, fast-developing and broadening responsibility, and status and income, the occupational safety and health director never had it so good.

What we need today is a man or woman who tells management what it could and should be doing, sells them on doing it and then gets the job done. We don't need followers or expedient or political operators in the field of environmental occupational protection; we need capable, tough-minded, innovative and progressive leadership, sensitive to and aware of the fact that both problem and solution is the worker on the line, who simply needs to know, understand and be trained in the job performed.

Life in this country is a business of individual elective. The environmental occupational protection director who can effectively translate that into a safe and healthful workplace and into safe and healthful work practices is at a premium today and will be more so as time passes.

The Act, essential as it is, is only a mechanism. Environmental occupational protection is a people problem, is people oriented and people produced and solved.

CHAPTER THIRTEEN
Safety Inspection Guide

Safety inspection today unfortunately has only one connotation—the arrival of the CSHO, the man from OSHA, who is going to make things difficult. This is indeed unfortunate because a safety inspection as suggested here was here long before the Act came into effect. It has been a profitable and valuable tool for industry employing it, and with or without an Act it is in some ways the principal key to a successful environmental occupational protection program.

If a general check were made regularly—checking specifics against the Standards to make certain of compliance—the impending visit of a CSHO would not be cause for alarm but rather an opportunity to check all areas with the CSHO with a minimum or absence of alleged violations and suggested fines. When the safety inspection guide is used in checking your plant, it is a good idea to make a list of those things that need to be done. Then waste no time in getting them done. One Safety Director made an inspection of one of his company's facilities, noting 83 violations of good safe practice. When the plant was visited later by a CSHO, there were three alleged violations, and one very minor fine was assessed.

HOW TO MAKE A SAFETY INSPECTION

DETECTING UNSAFE PRACTICES, CONDITIONS AND HAZARDS

To be effective, safety and fire protection activities have to be a continuing day-to-day part of supervision and not something that is merely thought of once a month at the time of safety meetings or a safety and fire protection inspection. Supervisors must be constantly alert to the physical conditions of their area in order to detect and correct unsafe conditions and fire hazards. When going through their area, they should look about them in order to observe the working habits and practices of their employees. Safe working habits are developed in employees when supervisors continuously demonstrate their interest by maintaining a firm and consistent approach and by taking immediate corrective action whenever an unsafe practice is observed. In this manner, employees soon realize that they are expected to work safely at all times and will thus develop good working habits. Some of the main points are itemized below:

1. The basic responsibility for maintaining a safe environment rests with the supervisor of the area. Periodic inspections by supervision are, therefore, essential. It is desirable that a member of supervision from the area being inspected accompany any outside inspector to explain applicable area operations rules, and to exercise close surveillance in detecting unsafe practices and conditions which have not been recognized during day to day activities.

2. Outside inspectors should report to area supervisors and explain in detail what violations were found, if area supervisors did not accompany them during the inspection.

3. An inspector should remember that his conduct must be above reproach while inspecting an area. To do this he must be familiar with and obey all area safety rules, such as the wearing of respirators, hard hats, eye protection, special shoes, long sleeves, etc. Nothing makes so bad an impression as a safety inspector violating area rules.

4. When making a report say the same things to all concerned. For example, do not tell an area supervisor that his area is good and write a report to his boss about how the area is falling apart. Tell the same story to all levels of supervision.

5. When reporting unsafe practices and conditions, it is more important to report the cause than the item itself. Unless the cause is corrected the item will continue to appear on all future inspections.

6. Follow-up. An inspection that is not followed up to check on results is only half done. Make sure that you follow up on the action taken by area supervision. Promptness and sincerity of purpose should be displayed in the course of correcting unsafe conditions and situations.

7. Another point of caution is that supervision cannot look the other

way or condone relatively minor unsafe acts to avoid an unpleasant situation. Such action is detrimental to the entire safety program and will undermine the morale of employees.

MAIN POINTS TO LOOK FOR

Observe unsafe practices and conditions and make corrective action through line supervision. However, if an unsafe condition or practice could result in an injury at that particular time, the job should be stopped immediately and corrective action taken. An unsafe practice is defined as the breaking of any safety or operating rule, or any other act performed by an individual or individuals in which injury could result.

A few of the common points to be observed are itemized below:

UNSAFE CONDITIONS
1. missing guards on machinery, valves, flanges, etc.
2. tripping and slipping hazards
3. cluttered workspace
4. broken glass
5. cutting and pinching hazards
6. improper lock-out procedure
7. inspection on ladders delinquent
8. inspection on electric cords delinquent
9. inspection on steam hoses delinquent
10. improper use of safety rope
11. compressed gas cylinders not anchored, labeled or stored correctly
12. unguarded electric transmission lines in vicinity of men working
13. leaking process and service lines
14. safety chains
15. missing vapor proof globes on lights
16. electric control switches not identified
17. safety equipment in poor condition
18. uninsulated heat sources (steam lines, heat exchangers, etc.)
19. dusts, fumes, gases and vapors
20. lack of proper warning signs where needed

UNSAFE PRACTICES
1. climbing on tank tops, pipe lines, etc. without proper safety equipment
2. failure to wear proper personal protective equipment such as eye protection, hard hats, long sleeves, etc.

3. unauthorized smoking
4. working from ladders not tied off or held
5. working without proper and applicable permits
6. working on jobs without first having locked-out correctly
7. using steam hoses with pressure above 25 p.s.i. without proper authorization
8. failure to observe plant traffic regulations
9. poor pedestrian practices
10. working with improper tools

REFERENCE SECTION

Specific measurements referred to in this section are to be used for new installations. Although these may be desirable for existing installations, it must be pointed out that these measurements represent ideal conditions and judgment should be exercised as to whether or not existing facilities should be altered. Existing locations not meeting these requirements should be properly marked to identify any existing hazard. This section is primarily intended for reference.

BUILDINGS AND GROUNDS

RAMPS
1. adequate width and evenness for safe operation
2. not too steep for safe operation and never over 20 degrees unless auxiliary means for lowering
3. free of slipping and tripping hazards
4. surfaced with suitable nonslip material whenever danger of slipping is present
5. ramps less than 4 feet above the ground shall not require handrails

OVERHEAD MINIMUM CLEARANCES
1. platforms, stairs, and other regular routes of foot travel: 7′ 0″
2. roadways: 15′ 0″
3. broad-gauge tracks: 22′ 0″

LATERAL MINIMUM CLEARANCES
1. regular routes of foot travel: 36″
2. passage between machines: 24″
3. safety exits and doorways: 28″
4. gangplanks and runways: 30″
5. runways for two-wheel hand trucks (two-way traffic): 60″
6. broad-gauge tracks: 8′ 0″ from center of track

WINDOWS
1. does pivot sash create bumping hazard?
2. does cracked or broken glass create hazard?
3. is sash in poor repair or in need of paint?
4. do latches work? are safety chains in place?
5. check condition of sash cords, chains or tubular balances (for fire windows, see Fire Protection Equipment)

VENTILATION
1. are ventilated hoods regularly inspected for air velocity?

DOORS
1. height of doors shall be 7' 0" or taller. Interior doors with light traffic and no overhead obstructions may be 6' 8"
2. double swinging doors have transparent panel
3. should not swing toward stairs unless adequate platform is provided
4. open outward where practical

DRAINS AND DITCHES
1. are they kept clear of obstructions?
2. do they have fume and vapor traps where needed?
3. are they located for maximum drainage?
4. are they properly covered to prevent tripping?
5. are they properly guarded?

CAR SPOTS
1. ease of operation
2. static ground wires maintained
3. proper storage of measuring rods, sample bottles. Protectoseal dome covers, gaskets, flanges, car chocks, signs, etc.
4. adequate lighting for night work
5. safety shower for corrosive materials
6. indicating lights for retractable platform working (if present)

EXITS
1. buildings two stories or more—a minimum of two exits
2. exits must never be blocked
3. emergency exit doors—use only safety latches
4. safety chute: good repair, no obstructions, not for material slide
5. lights provided over each exit; and on at night if building is used
6. doors open outward

ELECTRICAL EQUIPMENT
1. fuses and wire of proper size (not over 15 amps on lighting circuits)
2. no missing conduit covers
3. loose or missing grounds
4. 3-wire or self-insulating cords on all portable devices
5. all devices approved for electrical classification of building (check electric clocks, telephones, desk lamps, motors, switches, etc.)
6. flashlights: only explosion-proof allowed for plant use, unless such are not necessary
7. proper identification on switches and controls

ILLUMINATION
1. adequate for type of work
2. clean globes free of cracks
3. proper gaskets under globe
4. use proper type and size of globe
5. burned-out lamps promptly replaced

STAIRS
1. risers 9" high maximum
2. tread width 10" minimum exclusive of 1" nosing
3. stairs carrying 50 or more persons at one time at least 44" wide, those carrying 10-50—36" wide, and those carrying less than 10 people—30"
4. landing width to be same as stair widths
5. railing 39" above outer edge of tread excluding nosing
6. stairs 4' or wider or space between side of stairs and wall: double railing. Handrails required on both sides of stairs 44" or wider
7. keep clear at all times

RAILINGS
1. not less than 30" above floor; midtail 21" above floor (4" toeboard where necessary) at all open sides and ends of platforms if vertical drop is 4' or more
2. required for lesser heights if conditions warrant, i.e., over moving machinery, open tanks, etc.

FLOORS
1. smooth for walking or trucking
2. adequate support for stored material
3. marked storage areas

4. clear of tripping and slipping hazards
5. ramps or steps over low obstructions
6. holes properly guarded

OUTSIDE WALKS
1. level and clear of obstructions
2. at least 3' wide
3. adequate drainage

ELEVATORS AND LIFTS
1. inspected once per month
2. pits kept clean and drained
3. door interlocks function
4. operated only by authorized personnel
5. load sign posted

HOISTS, CHAIN BLOCKS, WINCHES, ETC.
1. visually inspected quarterly and an annual teardown inspection; year of teardown stenciled on equipment
2. personnel not under loads
3. never overload—capacity marked on hoist
4. any evidence of damage to any part
5. adequate supporting steel

HEATING AND/OR AIR CONDITIONING
1. exposed steam lines insulated up to 6' 6" above floor level
2. steam pressure to radiators not to exceed 25 p.s.i.; to blowers 40 p.s.i.
3. all equipment in good condition

FILE CABINET AND DRAWERS
1. file cabinets anchored to prevent tilting (two or more may be bolted together)
2. drawer stops provided

PORTABLE ELECTRIC EQUIPMENT AND EXTENSION CORDS
1. inspected monthly and tagged

NONPORTABLE ELECTRIC EQUIPMENT
1. inspected annually and tagged, except on instruments included as part of the preventive maintenance program

BONDING
1. painted red in "odd" years and yellow in "even" years; bonded with nylon string

ELECTRICAL AND STATIC GROUNDING
1. on storage tanks
2. on tank car loading platforms
3. on transformer bank fences
4. on flammable liquid containers
5. on pipe lines entering hazardous areas or buildings
6. on tank trucks
7. on tank cars and railroad tracks
8. on electric motors
9. on steam or flammable liquid hoses
10. on explosives buildings
11. for lightning protection

SAFETY EQUIPMENT

SAFETY SHOWERS
1. test once per shift until water is clear, or mud is out
2. during freezing weather, make certain that the unlagged portion of the shower drains properly and that there is no ice formation on the floor or ground under the shower
3. on steam-traced safety showers, make sure first water that comes out is not hot
4. do not obstruct approach to shower
5. valve handle should pull toward person using shower and operate easily
6. green light should always be on, with switch properly marked to stay on
7. valve handle placed 3' above floor level and located where needed

STRETCHER BOXES
1. know location
2. check stretcher hinges
3. check box hinges and latch
4. check points where canvas is attached to wood frames

AIR MASKS
1. maintained
2. check for proper use and connections in field. Valves blocked open by man wearing air mask, prepared and tagged by qualified E.R.D. attendant

GAS MASKS
1. stored in labeled mask box; latch and hinges on box work easily; no other materials stored in box

2. sealed in plastic bag (does not apply to Chemox or Scott)
3. inspected monthly
4. box properly located for emergency use
5. return used masks for service
6. canister gas masks acceptable only for protection against nuisance concentrations of toxic materials or for use as a high grade respirator

LIFE BELTS AND ROPES
1. tested every six months and marked
2. stored neatly in properly labeled box
3. no other materials stored in box
4. rope or belt should have no visual evidence of corrosion or contamination

ACID SUITS
1. stored properly
2. kept clean
3. undamaged
4. clear lens for hood

GOGGLE CLEANING STATION
1. adequate supply of cleaning tissues
2. receptacle for disposal of tissue
3. evidence that it is being used

SAFETY RULES
1. properly posted and up to date
2. cover new processes
3. reviewed periodically
4. covered with all new or transferred employees

BULLETIN BOARD
1. neat display
2. no obsolete (outdated) material or notices
3. good location

SAFETY SIGNS
1. right signs for job
2. signs in good condition

SAFETY ROPE
1. used to temporarily rope off unsafe conditions
2. properly tagged
3. used in proper locations
4. removed promptly after unsafe condition has been eliminated

EYE WASH BOTTLES
1. full and in good condition
2. located near applicable operations

FIRE PROTECTION EQUIPMENT

FIRE DOORS
1. not obstructed in any way
2. closes completely
3. counterweight is guarded
4. door is in good condition
5. door is printed red
6. rollers, track and chain are properly lubricated

FIRE ALARM BOXES
1. know location
2. unobstructed
3. red light always on
4. know how to operate
5. know how to report a fire

FIRE HYDRANT AND HOSE STATIONS AND HOUSES
1. know locations of all boxes within area
2. is hose dry and clean
3. nozzles and wrenches O.K.
4. box in good repair
5. access not blocked

FIRE EXTINGUISHERS
1. unobstructed
2. proper type of extinguisher for potential fires
3. post on background marked with red paint on all extinguishers (office buildings excepted)
4. check hose for plugged nozzle tips
5. condition of hose, especially where it is connected to extinguisher
6. lift or look to see if it is full
7. check last inspection date (annually)
8. is location best—near exit and fire source
9. extinguisher should be numbered

SPRINKLER SYSTEM
1. does it give complete coverage (under tanks, platforms, roof peaks, main building supports, flammable drumming booth, dryers).
2. nothing stored within 18" of sprinkler heads
3. nothing hung from sprinkler pipes

4. heads not corroded or painted over
5. lines identified by good paint job (red in operating building)
6. post indicator valve open
7. not to be used for electrical ground

MATERIAL HANDLING

STORAGE AND PILING
1. neat and orderly
2. not too high
3. accessible
4. personnel wearing proper protective equipment
5. proper identification
6. safe mechanical devices
7. loads not too bulky or heavy for employees or mechanical equipment
8. operators of mechanical equipment properly trained

PRECAUTIONS TAKEN FOR FOLLOWING
1. heavy
2. long
3. rough, sharp
4. slippery
5. flammable
6. hot
7. cold
8. fragile
9. poisonous
10. corrosive

OPERATING EQUIPMENT

PIPE LINES
1. flange and valve guards where needed
2. properly hung
3. remove if obsolete
4. safety valves where required

GAS CYLINDERS
1. transporting (damaged cap threads? regulators off? protective cap in place?)
2. storage (upright with chain)
3. supporting
4. manifolding (supplied by vendor, not home made; don't force odd threads)
5. identification (properly labeled; tear off tag properly used)

GUARDS
1. does it adequately cover moving machinery?
2. is it a permanent guard?
3. check condition carefully

GRINDING WHEELS
1. check for Perks washers (flanges); (cover at least 50% of wheel)
2. check position of tool rest (not to exceed 1/8 inch from wheel)
3. check guards around wheel

LAYOUT
1. adequate room for moving materials
2. adequate passageways and exits

TOOLS
1. inspected every three months
2. properly stored
3. in good condition
4. right tool for job

MAINTENANCE
1. inspect overall physical condition, paint, etc. of building and equipment
2. housekeeping

VENTS AND FLAME ARRESTORS
1. vessels containing flammable materials properly protected
2. each unit inspected quarterly

CONTROLS
1. properly identified

TEMPORARY EQUIPMENT
1. safely installed
2. adequate for job
3. not to be used for other purposes
4. should be replaced by permanent installation

LADDERS
1. portable ladders inspected semi-annually
2. fixed steel ladders inspected annually and marked
3. tie rope
4. check for damage
5. proper feet (Dayton A-500 ladder feet or spikes)
6. proper storage

STEAM HOSES
1. identified by yellow painted section on both ends
2. inspected quarterly by pipe shop and tagged
3. equipped with Boss coupling

AIR HOSES
1. inspected quarterly by pipe shop and tagged

PRESSURE RELIEF VALVE AND RUPTURE DISCS
1. applicable installations properly sized
2. inspected annually (month and year on lead seal)

UNFIRED PRESSURE VESSELS
1. all vessels qualifying as unfired pressure vessels must be hydrostatically tested at least once every five years

SCAFFOLDS
1. scaffold planks not to be used for any other purpose; inspected annually and marked with year inspected
2. scaffolds built correctly and inspected each day by those people using them

HOUSEKEEPING
1. neat and orderly arrangement of process material and portable equipment
2. floors and equipment clean
3. no foreign materials
4. a place for everything and everything in its place
5. adequate trash disposal facilities being used
6. facilities for disposal of dangerous materials such as:
    a. acids or other irritating chemicals
    b. explosives
    c. oxidizing agents
    d. glass or other sharp objects
    e. other applicable items not listed above

SANITATION AND HEALTH

TOILETS AND URINALS
1. keep clean
2. adequate supply of tissue
3. adequate ventilation

WASHING AND BATHING FACILITIES
1. hands and face washed before lunch when working with toxic materials

2. shower after exposure to contamination and/or at end of work shift
3. adequate hot water, basins, and number of showers to take care of rush periods
4. enough soap, other cleansers, paper towels
5. facilities kept clean
6. water temperature at faucet not over 145°F.

DRINKING WATER
1. only from drinking water lines
2. no connections between drinking water and any other lines
3. convenient locations
4. kept cool
5. fountains clean

LOCKERS
1. adequate storage space
2. adequate ventilation
3. periodic inspections
4. properly anchored lockers
5. benches provided

EATING
1. where permitted
2. a convenient, ventilated lunch room separated from operating area
3. kept clean
4. proper trash disposal facilities

PERSONNEL

CLOTHING
1. free of contaminants
2. no frayed edges to trip or catch in equipment
3. flame retardant where required
4. adequate coverage: sleeves buttoned at wrist, one button open at neck in area required
5. hat or cap
6. safety shoes in good condition

PERSONAL PROTECTIVE EQUIPMENT
1. safety glasses, properly fitted
2. gloves for type of work
3. hard hats required

4. availability and use of goggles, face masks, safety belts, air line respirators, acid suits, etc. where needed

UNSAFE PRACTICES
1. observe unsafe practices and conditions and make corrective action through line supervision. However, if an unsafe condition or practice could result in an injury at that particular time, the job should be stopped immediately and corrective action taken

SAFETY MEETINGS
1. held at regular intervals (at least monthly)
2. held on time
3. planned
4. effective
5. cover entire group

JOB TRAINING
1. new men should be well instructed in their job procedures
2. older employees should be retrained for new job
3. job procedures should be up-to-date and readily available

CHAPTER FOURTEEN
National Personnel Administrating The Occupational Safety and Health Act of 1970

DEPARTMENT OF LABOR

Most offices of the Department of Labor are located in the Main Labor Building at Third Street and Constitution Avenue, N.W., Washington, D.C. 20210, and offices listed in this directory, unless otherwise noted, are in that building.

Secretary of Labor .... W. J. Usery
523-8271

Assistant Secretary of Labor for
    Occupational Safety and Health .... Morton Corn
523-9362

Deputy Assistant Secretary for
    Occupational Safety and Health .... Bert M. Conklin
523-6091

OCCUPATIONAL SAFETY AND HEALTH ADMINISTRATION REGIONAL, AREA AND DISTRICT OFFICES; NAMES OF COORDINATING PERSONNEL

REGION I — Boston
(Connecticut, Maine, Massachusetts, New Hampshire, Rhode Island, Vermont)

Regional Administrator: Vacant, U.S. Department of Labor, Occupational Safety & Health Administration, 400-2 Totten Pond Road, Boston, Massachusetts 02154. FTS & Comm. Telephone: 617-223-6712/3.

AREA OFFICES

Francis R. Amirault, Area Director, U.S. Department of Labor, Occupational Safety & Health Administration, Federal Building, Room 426, 55 Pleasant Street, Concord, New Hampshire 03301. Comm. Telephone: 603-224-1995. FTS Telephone: 834-4725/4785.[1]

Rudolph Bayerle, Jr., Area Director, U.S. Department of Labor, Occupational Safety & Health Administration, U.S. Post Office and Courthouse Building, Room 501, 436 Dwight Street, Springfield, Massachusetts 01103. Comm. Telephone: 413-781-2420 Ext. 522. FTS Telephone: 836-9522/3.

John V. Fiatarone, Area Director, U.S. Department of Labor, Occupational Safety & Health Administration, Custom House Building, Room 703, State Street, Boston, Massachusetts 02109. FTS & Comm. Telephone: 617-223-4511/2.

Harold R. Smith, Area Director, U.S. Department of Labor, Occupational Safety & Health Administration, Federal Building, Room 617B, 450 Main Street, Hartford, Connecticut 06103. FTS & Comm. Telephone: 203-244-2294.

District Office

U.S. Department of Labor, Occupational Safety & Health Administration, Federal Building, Room 503A, U.S. Courthouse, Providence, Rhode Island 02903. Comm Telephone: 401-528-4466. FTS Telephone: 838-4466.

---

[1] FTS phone numbers are used by federal government employees, and commercial phone numbers are used by the general public. When there is direct dialing, as for the Boston Regional Office, one number is used for both purposes. In other cases, as for the Concord Area Office, separate numbers are used.

REGION II — New York
(New Jersey, New York, Puerto Rico, Virgin Islands)

Alfred Barden, Regional Administrator, U.S. Department of Labor, Occupational Safety & Health Administration, 1515 Broadway (1 Astor Plaza), Room 3445, New York, New York 10036. Comm. Telephone: 212-971-5941. FTS Telephone: 660-5941.

AREA OFFICES

Harry Allendorf, Area Director, U.S. Department of Labor, Occupational Safety & Health Administration, 519 Federal Street, Room 408, Camden, New Jersey 08101. Comm. Telephone: 609-757-5181. FTS Telephone: 488-5181.

Robert Barrett, Area Director, U.S. Department of Labor, Occupational Safety & Health Administration, 200 Mamaroneck Avenue, Room 302, White Plains, New York 10601. Comm. Telephone: 914-761-4250 Ext. 721. FTS Telephone: 883-4721.

David Bernard, Area Director, U.S. Department of Labor, Occupational Safety & Health Administration, Leo W. O'Brien Federal Building, Clinton Avenue & North Pearl Street, Room 132, Albany, New York 12207. Comm. Telephone: 518-472-6085. FTS Telephone: 562-6085.

James Conlon, Area Director, U.S. Department of Labor, Occupational Safety & Health Administration, Building T3, Belle Mead GSA Depot, Belle Mead, New Jersey 08502. Comm. Telephone 201-359-2777.

Nicholas DiArchangel, Area Director, U.S. Department of Labor, Occupational Safety & Health Administration, 90 Church Street, Room 1405, New York, New York 10007. FTS & Comm. Telephone: 212-264-9840.

Nicholas DiArchangel, Acting Area Director, U.S. Department of Labor, Occupational Safety & Health Administration, 271 Cadman Plaza, Room 629, Brooklyn, New York 11201. Comm. Telephone: 212-596-3440. FTS Telephone: 666-3440.

William Dreeland, Acting Area Director, U.S. Department of Labor, Occupational Safety & Health Administration, Teterboro Airport Professional Bldg., 377 Route 17, Room 206, Hasbrouck Heights, New Jersey 07604. Comm. Telephone: 201-288-1700.

Francisco Encarnacion-Rosa, Area Director, U.S. Department of Labor, Occupational Safety & Health Administration, Condominium San Alberto Bldg., 605 Condado Avenue, Room 328, Santurce, Puerto Rico 00907. Comm. Telephone: 809-724-1059.

James Epps, Area Director, U.S. Department of Labor, Occupational Safety & Health Administration, 370 Old Country Road, Garden City, L.I., New York 11530. Comm. Telephone: 516-294-0400.

James Epps, Acting Area Director, U.S. Department of Labor, Occupational Safety & Health Administration, 136-21 Roosevelt Avenue, Flushing, New York 11354. Comm. Telephone: 212-445-5005.

Charles Meister, Area Director, U.S. Department of Labor, Occupational Safety & Health Administration, 970 Broad Street, Room 1435 C, Newark, New Jersey 07102. Comm. Telephone: 201-645-5930. FTS Telephone: 341-5930.

Richard Palmieri, Area Director, U.S. Department of Labor, Occupational Safety & Health Administration, 2E Blackwell Street, Dover, New Jersey 07801. Comm. Telephone: 201-361-4050.

P. Charles Schwender, Area Director, U.S. Department of Labor, Occupational Safety & Health Administration, 111 W. Huron Street, Room 1002, Buffalo, New York 14202. Comm. Telephone: 716-842-3333. FTS Telephone: 432-3333.

Chester Whiteside, Area Director, U.S. Department of Labor, Occupational Safety & Health Administration, Midtown Plaza, Room 203, 700 E. Water Street, Syracuse, New York 13210. Comm Telephone: 315-473-2700. FTS Telephone: 951-2700.

(Vacant). U.S. Department of Labor, Occupational Safety & Health Administration, Federal Office Bldg., Room 600, 100 State Street, Rochester, New York 14614. Comm. Telephone: 716-263-6755. FTS Telephone: 473-6755.

REGION III — Philadelphia

(Delaware, District of Columbia, Maryland, Pennsylvania, Virginia, West Virginia)

David H. Rhone, Regional Administrator, U.S. Department of Labor, Occupational Safety & Health Administration, Gateway Building, Suite 15220, 3535 Market Street, Philadelphia, Pennsylvania 19104. Comm. Telephone: 215-596-1201.

AREA OFFICES

W. Gary Adams, Acting Area Director, U.S. Department of Labor, Occupational Safety & Health Administration, Charleston National Plaza, Room 1726, 700 Virginia Street, Charleston, West Virginia 25301. Comm. Telephone: 304-343-6181 Ext. 420/429. FTS Telephone: 924-1420.

Byron R. Chadwick, Area Director, U.S. Department of Labor, Occupational Safety & Health Administration, Federal Building, Room 1110A, Charles Center, 31 Hopkins Plaza, Baltimore, Maryland 21201. Comm. Telephone: 301-962-2840. FTS Telephone: 920-2840.

Lapsley Ewing, Jr., Area Director, U.S. Department of Labor, Occupational Safety & Health Administration, Federal Building (P.O. Box 10186), Room 8018, 400 North 8th Street, Richmond, Virginia 23240. Comm. Telephone: 804-782-2864/5. FTS Telephone: 925-2864/5.

Robert Farranato, Safety Specialist, U.S. Department of Labor, Occupational Safety & Health Administration, Penn Place, Room 3107, 20 N. Pennsylvania Avenue, Wilkes-Barre, Pennsylvania 18701. Comm. Telephone: 717-825-6811 Ext. 538/9. FTS Telephone: 592-6811 Ext. 538/9.

Harry Lacey, Area Director, U.S. Department of Labor, Occupational Safety & Health Administration, Jonnet Building, Room 802, 4099 William Penn Highway, Monroeville, Pennsylvania 15146. Comm. Telephone: 412-644-2905. FTS Telephone: 722-2905.

Harry Sachkar, Area Director, U.S. Department of Labor, Occupational Safety & Health Administration, William J. Green, Jr. Federal Building, Room 4256, 600 Arch Street, Philadelphia, Pennsylvania 19106. FTS & Comm. Telephone: 215-597-4955.

Gilbert Soulter, Acting Area Director, U.S. Department of Labor, Occupational Safety & Health Administration, Railway Labor Building, Room LL-2, 400 First Street, N.W., Washington, D.C. 20210. Comm. Telephone: 202-961-5132.

Progress Plaza, 49 N. Progress Avenue, Harrisburg, Pennsylvania 17109.

District Office

Charles Straw, Safety Specialist, U.S. Department of Labor, Occupational Safety & Health Administration, Stanwick Building, Room 111, 3661 Virginia Beach Blvd., Norfolk, Virginia 23502. Comm. Telephone: 804-441-6381. FTS Telephone: 939-6381.

Field Stations

Farris Anderson, Safety Specialist, U.S. Department of Labor, Occupational Safety & Health Administration, Carlton Terrace Building, Suite 114, 920 South Jefferson Street, Roanoke, Virginia 24016. Comm. Telephone: 703-343-1581 Ext. 271. FTS Telephone: 937-6271.

Laurence Barker, Safety Specialist, U.S. Department of Labor, Occupational Safety & Health Administration, Armenara Office Center, Suite 470, 444 E. College Avenue, State College, Pennsylvania 16801. Comm. Telephone: 814-234-6695.

George Cantwell, Safety Specialist, U.S. Department of Labor, Occupational Safety & Health Administration, Falls Church Office Building, Room 107, 900 South Washington Street, Falls Church, Virginia 22046. Comm. Telephone: 703-557-1330.

Charles Green, Safety Specialist, U.S. Department of Labor, Occupational Safety & Health Administration, Central Plaza, Room B, East King Street, Lancaster, Pennsylvania 17602. Comm. Telephone: 717-394-0681 Ext. 73 or 74. FTS Telephone: 592-0573.

Alonzo L. Griffin, Safety Engineer, U.S. Department of Labor, Occupational Safety & Health Administration, Federal Office Building, Room 3007, 844 King Street, Wilmington, Delaware 19801. Comm. Telephone: 302-571-6115. FTS Telephone: 487-6115.

James Troy, Safety Specialist, U.S. Department of Labor, Occupational Safety & Health Administration, U.S. Court House, Room 411, Chapline and 12th Streets, Wheeling, West Virginia 26003. Comm. Telephone: 304-232-8044. FTS Telephone: 923-1062.

Clair Waltz, Safety Specialist, U.S. Department of Labor, Occupational Safety & Health Administration, Burneson Building, 933 Park Avenue, Meadville, Pennsylvania 16335. Comm. Telephone: 814-724-8031. FTS Telephone: 722-8031.

U.S. Department of Labor, Occupational Safety & Health Administration, U.S. Post Office Bldg., Room 22, 5th & Hamilton Streets, Allentown, Pennsylvania 18101. Comm. Telephone: 215-434-0181 Ext. 266. FTS Telephone: 488-0181.

REGION IV — Atlanta

(Alabama, Florida, Georgia, Kentucky, Mississippi, North Carolina, South Carolina, Tennessee)

Donald E. Mackenzie, Regional Administrator, U.S. Department of Labor, Occupational Safety & Health Administration, 1375 Peachtree Street, N.E., Suite 587, Atlanta, Georgia 30309. Comm. Telephone: 404-526-2281/3573. FTS Telephone: 285-3573.

AREA OFFICES

George Barlow, Area Director, U.S. Department of Labor, Occupational Safety & Health Administration, Bridge Building, Room 204, 3200 E. Oakland Park Boulevard, Fort Lauderdale, Florida 33308. Comm. Telephone: 305-350-9331. FTS Telephone: 565-4211.

James Blount, Area Director, U.S. Department of Labor, Occupational Safety & Health Administration, 5760-I-55 North Frontage Road East, Jackson, Mississippi 39211. Comm. Telephone: 601-969-4606. FTS Telephone: 490-4606.

Joseph Camp, Area Director, U.S. Department of Labor, **Occupational** Safety & Health Administration, Building 10, Suite 33, La Vista Perimeter Office Park, Tucker, Georgia 30084. Comm. Telephone: 404-939-8987. FTS Telephone: 285-4767.

Raymond G. Finney, Area Director, U.S. Department of Labor, Occupational Safety & Health Administration, 1710 Gervais Street, Room 205, Columbia, South Carolina 29201. Comm. Telephone: 803-765-5904. FTS Telephone: 677-5904.

Frank Flanagan, Area Director, U.S. Department of Labor, Occupational Safety & Health Administration, 600 Federal Place, Suite 554-E, Louisville, Kentucky 40202. Comm. Telephone: 502-582-6111/2. FTS Telephone: 352-6111/2.

William Gordon, Area Director, U. S. Department of Labor, Occupational Safety & Health Administration, Art Museum Plaza, Suite 4, 2809 Art Museum Drive, Jacksonville, Florida 32207. Comm. Telephone: 904-791-2895. FTS Telephone: 946-2895.

Quinton Haskins, Area Director, U.S. Department of Labor, Occupational Safety & Health Administration. Federal Office Building, Room 406, 310 New Bern Avenue, Raleigh, North Carolina 27601. Comm. Telephone: 919-755-4770. FTS Telephone: 672-4770.

A. DeJean King, Area Director, U.S. Department of Labor, Occupational Safety & Health Administration, Riverside Plaza Shopping Center, 2720 Riverside Drive, Macon, Georgia 31204. Comm. Telephone: 912-734-2264. FTS Telephone: 284-0264.

Eugene Light, Area Director, U.S. Department of Labor, Occupational Safety & Health Administration, Suite 302, 1600 Hayes Street, Nashville, Tennessee 37203. Comm. Telephone: 615-749-5313. FTS Telephone: 852-5313.

Harold Monegue, Area Director, U.S. Department of Labor, Occupational Safety & Health Administration, Commerce Building, Room 600, 118 North Royal Street, Mobile, Alabama 36602. Comm. Telephone: 205-690-2131. FTS Telephone: 534-2131.

Edward G. Savage, Area Director, U.S. Department of Labor, Occupational Safety & Health Administration, Barnett Bank Bldg., Room 918, 1000 N. Ashley Drive, Tampa, Florida 33602. Comm. Telephone: 813-228-2821. FTS Telephone: 826-2821/23.

G. L. Wyatt, Area Director, U.S. Department of Labor, Occupational Safety & Health Administration, Todd Mall, 2047 Canyon Road, Birmingham, Alabama 35216. Comm. Telephone: 205-882-7100. FTS Telephone: 229-1541.

U.S. Department of Labor, Occupational Safety & Health Administration, Enterprise Building, Suite 210A, 6605 Abercorn Street, Savannah, Georgia 31405. Comm. Telephone: 912-354-0733. FTS Telephone: 287-4393.

Field Stations

Howard Gillingham, U.S. Department of Labor, Occupational Safety & Health Administration, Kozerama Building, 1300 Executive Center Drive, Tallahassee, Florida 32301. Comm. Telephone: 904-377-4286.

John Hall, U.S. Department of Labor, Occupational Safety & Health Administration, Aronov Building, Room 329, 474 South Court Street, Montgomery, Alabama 36103. Comm. Telephone: 205-832-7159. FTS Telephone: 534-7159.

Roy M. Hirano, U.S. Department of Labor, Occupational Safety & Health Administration, Post Office Box N-04, 129 Noble Street, Anniston, Alabama 36201. Comm. Telephone: 205-237-4212. FTS Telephone: 229-6951.

Willie H. Joiner, U.S. Department of Labor, Occupational Safety & Health Administration, 334 Meeting Street, Room 627, Federal Building, 6th Floor, Charleston, South Carolina 29403. Comm. Telephone: 803-577-2423. FTS Telephone: 667-4529.

Robert S. Krueger, U.S. Department of Labor, Occupational Safety & Health Administration, West Clinton Bldg., Room 122, 2109 Clinton Avenue West, Huntsville, Alabama 35805.

Kenneth Scarbrough, U.S. Department of Labor, Occupational Safety & Health Administration, P.O. Box 12212, Pensacola, Florida 32581. Comm. Telephone: 904-478-0830.

Larry K. Weaver, U.S. Department of Labor, Occupational Safety & Health Administration, U.S. Post Office Bldg., Room 204, Sheffield, Alabama 35660. Comm. Telephone: 205-383-0010.

Wayne Welborn, U.S. Department of Labor, Occupational Safety & Health Administration, Federal Building, Room 207, Memphis, Tennessee 38118. Comm. Telephone: 901-534-0179. FTS Telephone: 222-4179.

David L. Williams, U.S. Department of Labor, Occupational Safety & Health Administration, Murchinson Bldg., Room 613, 201 North Front Street, Wilmington, North Carolina 28401. Comm. Telephone: 919-791-6430. FTS Telephone: 674-9474.

REGION V — Chicago

(Illinois, Indiana, Michigan, Minnesota, Ohio, Wisconsin)

Edward E. Estkowski, Regional Administrator, U.S. Department of Labor, Occupational Safety & Health Administration, 32nd Floor, Room 3263, 230 South Dearborn Street, Chicago, Illinois 60604. FTS & Comm. Telephone: 312-353-4716/7.

AREA OFFICES

Kenneth Bowman, Area Director, U.S. Department of Labor, Occupational Safety & Health Administration, Federal Office Building, Room 847, 1240 East Ninth Street, Cleveland, Ohio 44199. Comm. Telephone: 216-522-3818. FTS Telephone: 293-3818.

Glenn Butler, Area Director, U.S. Department of Labor, Occupational Safety & Health Administration, Federal Office Building, Room 734, 234 North Summit Street, Toledo, Ohio 43604. Comm. Telephone: 419-259-7542. FTS Telephone: 625-7542.

Vernon Fern, Area Director, U.S. Department of Labor, Occupational Safety & Health Administration, 110 South Fourth Street, Room 437, Minneapolis, Minnesota 55401. FTS & Comm. Telephone: 612-725-2571.

William Funcheon, Safety & Health Manager, U.S. Department of Labor, Occupational Safety & Health Administration, 230 South Dearborn Street, 16th Floor, Chicago, Illinois 60604. FTS & Comm. Telephone: 312-353-1390.

Robert Hanna, Area Director, U.S. Department of Labor, Occupational Safety & Health Administration, Clark Building, Room 400, 633 West Wisconsin Avenue, Milwaukee, Wisconsin 53203. Comm. Telephone: 414-224-3315/6. FTS Telephone: 362-3315/6.

J. Fred Keppler, Area Director, U.S. Department of Labor, Occupational Safety & Health Administration, U.S. Post Office and Courthouse, Room 423, 46 East Ohio Street, Indianapolis, Indiana 46204. Comm. Telephone: 317-269-7290. FTS Telephone: 331-7290.

Ronald McCann, Area Director, U.S. Department of Labor, Occupational Safety & Health Administration, Federal Office Building, Room 4028, 550 Main Street, Cincinnati, Ohio 45202. FTS & Comm. Telephone: 513-684-2354.

Peter Schmitt, Area Director, U.S. Department of Labor, Occupational Safety & Health Administration, Room 109, 360 S. 3rd Street, Columbus, Ohio 43215. Comm. Telephone: 614-469-5582. FTS Telephone: 943-5582.

U.S. Department of Labor, Occupational Safety & Health Administration, Michigan Theatre Building, Room 626, 220 Bagley Avenue, Detroit, Michigan 48226. FTS & Comm. Telephone: 313-226-6720.

(Vacant), Area Director, U.S. Department of Labor, Occupational Safety & Health Administration, U.S. Postal Service Bldg., Room 204, 219 Washington Avenue, Oshkosh, Wisconsin 54901. Comm. Telephone: 414-231-1406/8. FTS Telephone: 352-0111 (FTS Operator).

U.S. Department of Labor, Occupational Safety & Health Administration, 228 N.E. Jefferson, 3rd Floor, Peoria, Illinois 61602. Comm. Telephone: 306-673-9515. FTS Telephone: 360-7033.

District Offices

U.S. Department of Labor, Occupational Safety & Health Administration, 305 S. Illinois Street, Belleville, Illinois 62220. Comm. Telephone: 618-277-5300.

U.S. Department of Labor, Occupational Safety & Health Administration, Federal Bldg., U.S. Courthouse, 500 Barstow Street, Room B-9, Eau Claire, Wisconsin 54701. Comm. Telephone: 715-832-9019. FTS Telephone: 832-2252.

U.S. Department of Labor, Occupational Safety & Health Administration, Office #3, 2326 S. Park Street, Madison, Wisconsin 53713.

U.S. Department of Labor, Occupational Safety & Health Administration, 317 First Street, Wausau, Wisconsin 54401. Comm. Telephone: 715-842-8004. FTS Telephone: 364-5000 (FTS Operator).

REGION VI — Dallas

(Arkansas, Louisiana, New Mexico, Oklahoma, Texas)

Robert Tice, Regional Administrator, U.S. Department of Labor, Occupational Safety & Health Administration, 555 Griffin Square Bldg., Room 602, Dallas, Texas 75202. FTS & Comm. Telephone: 214-749-2477/8/9.

AREA OFFICES

Thomas Curry, Area Director, U.S. Department of Labor, Occupational Safety & Health Administration, Riverview Professional Building, S. 77 Sunshine Strip, Suite 9, Harlingen, Texas 78550. Comm. Telephone: 512-425-6811/12. FTS Telephone: 734-4516/18.

Robert A. Griffin, Area Director, U.S. Department of Labor, Occupational Safety & Health Administration, 2320 Le Branch Street, Room 2118, Houston, Texas 77004. Comm. Telephone 713-226-5431. FTS Telephone: 527-5431.

W. E. Hargrove, Area Director, U.S. Department of Labor, Occupational Safety & Health Administration, Adolphus Towers, 1412 Main Street, Suite 1820, Dallas, Texas 75202. FTS & Comm. Telephone: 214-749-1786.

James Knorpp, Area Director, U.S. Department of Labor, Occupational Safety & Health Administration, Petroleum Building, Room 514, 420 South Boulder, Tulsa, Oklahoma 74103. Comm. Telephone: 918-581-7676. FTS Telephone: 736-7676.

Hubert M. Kurtz, Area Director, U.S. Department of Labor, Occupational Safety & Health Administration, Federal Courthouse Building, 200 West 8th Street, Room 9, Austin, Texas 78701. FTS & Comm. Telephone: 512-397-5783.

John Parsons, Area Director, U.S. Department of Labor, Occupational Safety & Health Administration, Federal Building, Room 3114, 500 Gold Avenue S.W., P.O. Box 1428, Albuquerque, New Mexico 87103. FTS & Comm. Telephone: 505-766-3411.

James E. Powell, Area Director, U.S. Department of Labor, Occupational Safety & Health Administration, 2156 Wooddale Boulevard, Hoover Annex, Suite 200, Baton Rouge, Louisiana 70806. Comm. Telephone: 504-387-0181 Ext. 474. FTS Telephone: 687-4474.

Marvin Schierman, Area Director, U.S. Department of Labor, Occupational Safety & Health Administration, 546 Carondelet Street, Room 202, New Orleans, Louisiana 70130. Comm. Telephone: 504-589-2451/2. FTS Telephone: 682-6166/7, 682-2451/2.

Robert Simmons, Area Director, U.S. Department of Labor, Occupational Safety & Health Administration, Federal Building, Room 421, 1205 Texas Avenue, Lubbock, Texas 79401. Comm. Telephone: 806-762-7681. FTS Telephone: 783-7681.

Lloyd A. Warren, Area Director, U.S. Department of Labor, Occupational Safety & Health Administration, 50 Penn Place, Oklahoma City, Oklahoma 73118. Comm. Telephone: 405-231-5351. FTS Telephone: 736-5351.

U.S. Department of Labor, Occupational Safety & Health Administration, Donaghey Building, Room 526, 103 East 7th Street, Little Rock, Arkansas 72201. Comm. Telephone: 501-378-6192. FTS Telephone: 740-6192.

District Office

Henry J. Ahlf, District Supervisor, U.S. Department of Labor, Occupational Safety & Health Administration, 600 Leopard Street, Suite 1322, Corpus Christi, Texas 78401. Comm. Telephone: 512-888-3257. FTS Telephone: 734- 3257.

Field Stations

Carlos Gonzales, Senior Compliance Officer, U.S. Department of Labor, Occupational Safety & Health Administration, 1515 Airway Blvd., Room 3, El Paso, Texas 79925. Comm. Telephone 915-543-7828. FTS Telephone: 572-7828.

Fred N. McKenzie, U.S. Department of Labor, Occupational Safety & Health Administration, New Federal Office Building, Room 8A09, 500 Fannin Street, Shreveport, Louisiana 71101. Comm. Telephone: 318-226-5360. FTS Telephone: 493-5360.

U.S. Department of Labor, Occupational Safety & Health Administration, Professional Building, Suite 300, 2900 North Street, Beaumont, Texas 77702. Comm. Telephone: 713-838-0271 Ext. 258/9. FTS Telephone: 527-2911.

U.S. Department of Labor, Occupational Safety & Health Administration, 1015 Jackson Keller Road, Room 215, San Antonio, Texas 78213. FTS & Comm. Telephone: 512-225-4569.

REGIONAL VII — Kansas City
(Iowa, Kansas, Missouri, Nebraska)

Vernon A. Strahm, Regional Administrator, U.S. Department of Labor, Occupational Safety & Health Administration, 911 Walnut Street, Room 3000, Kansas City, Missouri 64106. Comm. Telephone: 816-374-5861. FTS Telephone: 758-5861.

AREA OFFICES

Carmine Barone, Acting Area Director, U.S. Department of Labor, Occupational Safety & Health Administration, 113 West 6th Street, North Platte, Nebraska 69101. Comm. Telephone: 308-534-9450.

Robert Borchardt, Area Director, U.S. Department of Labor, Occupational Safety & Health Administration. 1627 Main Street, Room 1100, Kansas City, Missouri. Comm. Telephone 816-374-2756. FTS Telephone: 758-2756.

Angelo Castranova, Area Director, U.S. Department of Labor, Occupational Safety & Health Administration, 210 North 12th Boulevard, Room 554, St. Louis, Missouri 63101. Comm. Telephone: 314-425-5461. FTS Telephone: 279-5461.

Roger Clark, Area Director, U.S. Department of Labor, Occupational Safety & Health Administration, Petroleum Building, Suite 512, 221 South Broadway Street, Wichita, Kansas 67202. Comm. Telephone: 316-267-6311 Ext. 644. FTS Telephone: 752-6644.

Uldis Levalds, Area Director, U.S. Department of Labor, Occupational Safety & Health Administration, 210 Walnut Street, Room 638, Des Moines, Iowa 50309. Comm. Telephone: 515-284-4794. FTS Telephone: 862-4794.

Warren Wright, Area Director, U.S. Department of Labor, Occupational Safety & Health Administration, Room 100, Overland-Wolf Building, 6910 Pacific Street, Omaha, Nebraska 68106. Comm. Telephone: 402-221-3276. FTS Telephone: 864-3276.

REGION VIII — Denver

(Colorado, Montana, North Dakota, South Dakota, Utah, Wyoming)

Curtis Foster, Regional Administrator, U.S. Department of Labor, Occupational Safety & Health Administration, Federal Building, Room 15010, 1961 Stout Street, Denver, Colorado 80202. Comm. Telephone: 303-837-3883. FTS Telephone: 327-3883.

AREA OFFICES

Wallace Beebe, Jr., Acting Area Director, U.S. Department of Labor, Occupational Safety & Health Administration, Squire Plaza Building, 8527 West Colfax Avenue, Lakewood, Colorado 80215. FTS & Comm. Telephone: 303-234-4471.

Charles Hines, Area Director, U.S. Department of Labor, Occupational Safety & Health Administration, Court House Plaza Building, Room 408, 300 North Dakota Avenue, Sioux Falls, South Dakota 57102. Comm. Telephone: 605-336-2980 Ext. 425. FTS Telephone: 782-4425.

Harry Hutton, Area Director, U.S. Department of Labor, Occupational Safety & Health Administration, Petroleum Building, Suite 525, 2812 1st Avenue North, Billings, Montana 59101. Comm. Telephone: 406-245-6711 Ext. 6640/9. FTS Telephone: 585-6640.

Jerome Williams, Area Director, U.S. Department of Labor, Occupational Safety & Health Administration, Russel Building, Highway 83 North, Route 1, Bismarck, North Dakota 58501. Comm. Telephone: 701-255-4011 Ext. 521. FTS Telephone: 783-4521.

Ernest Yanni, Acting Area Director, U.S. Department of Labor, Occupational Safety & Health Administration, U.S. Post Office Building, Room 452, 350 South Main Street, Salt Lake City, Utah 84101. Comm. Telephone: 801-524-5080. FTS Telephone: 588-5080.

REGION IX — San Francisco

(Arizona, California, Hawaii, Nevada, Guam, American Samoa, Trust Territory of the Pacific Islands)

Gabriel Gillotti, Regional Administrator, U.S. Department of Labor, Occupational Safety & Health Administration, 9470 Federal Building, 450 Golden Gate Avenue, P.O. Box 36017, San Francisco, California 94102. FTS & Comm. Telephone: 415-556-0584.

AREA OFFICES

Gilbert Garcia, Area Director, U.S. Department of Labor, Occupational Safety & Health Administration, Amerco Towers, Suite 318, 2721 North Central Avenue, Phoenix, Arizona 85004. FTS & Comm. Telephone: 602-261-4858.

Paul Haygood, Area Director, U.S. Department of Labor, Occupational Safety & Health Administration, 333 Queen Street, Suite 505, Honolulu, Hawaii 96813. FTS & Comm. Telephone: 808-546-3157.

Ivan Schulenburg, Area Director, U.S. Department of Labor, Occupational Safety & Health Administration, 1100 East William Street, Suite 222, Carson City, Nevada 89701. Comm. Telephone: 702-883-1226.

Bernard Tibbetts, Area Director, U.S. Department of Labor, Occupational Safety & Health Administration, Hartwell Building, Room 401, 19 Pine Avenue, Long Beach, California 90802. Comm. Telephone: 213-548-2431. FTS Telephone: 796-2431.

U.S. Department of Labor, Occupational Safety & Health Administration, Room 1706, 100 McAllister Street, San Francisco, California 94102. FTS & Comm. Telephone: 415-556-7260.

Field Stations

Merle Annis, Safety Specialist, U.S. Department of Labor, Occupational Safety & Health Administration, 2110 Merced Street, Room 202, Fresno, California 93721. Comm. Telephone: 209-487-5454. FTS Telephone: 467-5454.

Donald Fischer, Safety Specialist, U.S. Department of Labor, Occupational Safety & Health Administration, Room 3-I, 301 W. Congress Street, Tucson, Arizona 85701. Comm. Telephone: 602-792-6286. FTS Telephone: 762-6286.

Dan Shipley, Safety Specialist, U.S. Department of Labor, Occupational Safety & Health Administration, 300 Las Vegas Blvd. South, Room I-606, Las Vegas, Nevada 89101. Comm. Telephone: 702-385-6570. FTS Telephone: 598-6570.

John Williams, Safety Specialist, U.S. Department of Labor, Occupational Safety & Health Administration, 2800 Cottage Way, Room 1409, Sacramento, California 95825. Comm. Telephone: 916-484-4363. FTS Telephone: 486-4363.

REGION X — Seattle
(Alaska, Idaho, Oregon, Washington)

James W. Lake, Regional Administrator, U.S. Department of Labor, Occupational Safety & Health Administration, Federal Office Bldg., Room 6048, 909 First Avenue, Seattle, Washington 98174. Comm. Telephone: 206-442-5930. FTS Telephone: 399-5930.

AREA OFFICES

Richard Beeston, Area Director, U.S. Department of Labor, Occupational Safety & Health Administration, 121-107th Street, N.E., Bellevue, Washington 98004. Comm. Telephone: 206-442-7520. FTS Telephone: 399-7520.

Eugene Harrower, Area Director, U.S. Department of Labor, Occupational Safety & Health Administration, Pittock Block, Room 526, 921 S.W. Washington Street, Portland, Oregon 97205. Comm. Telephone: 503-221-2251. FTS Telephone: 423-2251.

Richard Jackson, Area Director, U.S. Department of Labor, Occupational Safety & Health Administration, 1319 West Idaho Street, Boise, Idaho 83706. Comm. Telephone: 208-342-2711 Ext. 2867/8/9. FTS Telephone: 588-2867.

Ronald T. Tsunehara, Area Director, U.S. Department of Labor, Occupational Safety & Health Administration, Federal Building, Room 227, 605 West 4th Avenue, Anchorage, Alaska 99501.

Field Stations

U.S. Department of Labor, Occupational Safety & Health Administration, Room B17, Federal Building, U.S. Courthouse, Box 87, 150 South Arthur Street, Pocatello, Idaho 83201. Comm. Telephone: 208-233-6374. FTS Telephone: 588-6268.

U.S. Department of Labor, Occupational Safety & Health Administration, Room 410, 904 Riverside Avenue, U.S. Post Office, Box 2132, Spokane, Washington 99210. Comm. Telephone: 509-624-5235. FTS Telephone: 439-2598.

142 OSHA HANDBOOK

U.S. Department of Labor
Occupational Safety and Health Administration
March 1976

Region	Management Officers	AARD for Federal and State Operations	AARD for Training	AARD for Technical Support
Boston FTS Phone	Nicholos Pappas 223-6712	Edwin J. Riley, Jr. 223-6712	Edward C. Garvin 223-6712	Gerald M. DuWors 223-6712
New York FTS Phone	Jules Wise 660-5754	Richard F. Andree 660-7006	Carl Meyer 660-5921	John J. Kearney 660-5921
Philadelphia FTS Phone	John Moffa 596-1207	William Carrigan 596-1201	Nathan Richmond 596-1201	Nathan Richmond 596-1201
Atlanta FTS Phone	Alan McMillan 285-2285	Robert A. Wendell 285-3573	Richard Honey 285-3573	Cois M. Brown 285-3573
Chicago FTS Phone	Ron Leon 353-1366	Thomas Levenhagen 353-4318	Charles S. Wolf 353-1369	Edward Largent 353-1370
Dallas FTS Phone	Jack Brett 749-2477	Charles Freeman 749-2477	William Hargrove 749-2477	Clarence Holder 749-2477
Kansas City FTS Phone	Josephine Ferguson 758-5861	Harold L. Smith 758-5861	George Flexander 758-5861	Bernard D. Olson 758-5761
Denver FTS Phone	Donald Schworer 327-3883	Vacant	Vacant	Jack D. Torrey 327-3883
San Francisco FTS Phone	Hamilton Fairburn 556-7430	Ken Holland 556-1635	Hamilton Fairburn 556-7430	Donald Meyer 556-8743
Seattle FTS Phone	Steven McKinney 399-5930	L. Thomas Ashcraft 399-1080	Jack Jones 399-5930	John A. Granchi 399-5930

BUREAU OF LABOR STATISTICS REGIONAL OFFICES

REGION I

(Connecticut, Maine, Massachusetts, New Hampshire, Rhode Island, Vermont)

Bureau of Labor Statistics, 1603-A Federal Office Building, Boston, Massachusetts 02203. Telephone: 617-223-4541.

REGION II

(New Jersey, New York, Puerto Rico, Virgin Islands)

Bureau of Labor Statistics, 341 Ninth Avenue, New York, New York 10001. Telephone: 212-971-5915.

REGION III

(Delaware, District of Columbia, Maryland, Pennsylvania, Virginia, West Virginia)

Bureau of Labor Statistics, Penn Square Building, Room 406, 1317 Filbert Street, Philadelphia, Pennsylvania 19107. Telehone: 215-597-7510.

REGION IV

(Alabama, Florida, Georgia, Kentucky, Mississippi, North Carolina, South Carolina, Tennessee)

Bureau of Labor Statistics, 1317 Peachtree Street N.E., Atlanta, Georgia 30309. Telephone: 404-526-3660.

REGION V

(Illinois, Indiana, Michigan, Minnesota, Ohio, Wisconsin)

Bureau of Labor Statistics, 219 S. Dearborn Street, Chicago, Illinois 60604. Telephone: 312-353-7253.

REGION VI

(Arkansas, Louisiana, New Mexico, Oklahoma, Texas)

Bureau of Labor Statistics, 1100 Commerce Street, Room 6B7, Dallas, Texas 75202. Telephone: 214-749-1781.

REGIONS VII AND VIII

(VII—Iowa, Kansas, Missouri, Nebraska; VIII—Colorado, Montana, North Dakota, South Dakota, Utah, Wyoming)

Bureau of Labor Statistics, Federal Office Building, 911 Walnut Street, Kansas City, Missouri 64106. Telephone: 816-374-3685.

REGIONS IX AND X
(IX—Arizona, California, Hawaii, Nevada;
X—Alaska, Idaho, Oregon, Washington)

Bureau of Labor Statistics, 450 Golden Gate Avenue, Box 36017, San Francisco, California 94102. Telephone: 415-556-8980.

NATIONAL INSTITUTE FOR OCCUPATIONAL SAFETY AND HEALTH

Headquarters: Department of Health, Education and Welfare, Public Health Service, Center for Disease Control, NIOSH, Health Services and Mental Health Administration, Rockville, Maryland 20852.

Cincinnati Laboratory: NIOSH, R. A. Taft Sanitary Engineering Center, 4676 Columbia Parkway, Cincinnati, Ohio 45226.

Appalachian Laboratory: Appalachian Center for Occupational Safety and Health, NIOSH, P.O. Box 4292, Morgantown, West Virginia 26505.

Director: Dr. John Finklea; A. C. Blackman, Assistant Director for Safety.

Regional Directors and Offices:

REGION I — Boston
(Connecticut, Maine, Massachusetts, New Hampshire, Rhode Island, Vermont)

Regional Program Director, Paul Alvarado, NIOSH, DHEW Region I, Government Center (JFK Fed. Bldg.), Boston Massachusetts 02203.

REGION II — New York City
(New Jersey, New York, Puerto Rico, Virgin Islands)

Regional Program Director, Mary Louise Brown, RN, NIOSH, DHEW Region II, Federal Bldg., 26 Federal Plaza, New York, New York 10007.

REGION III — Philadelphia
(Delaware, District of Columbia, Maryland, Pennsylvania, Virginia, West Virginia)

Regional Program Director, William Shoemaker, NIOSH, DHEW Region III, 401 North Broad Street, Philadelphia, Pennsylvania 19108.

REGION IV — Atlanta
(Alabama, Florida, Georgia, Kentucky, Mississippi, North Carolina, South Carolina, Tennessee)

Regional Program Director, Gordon Nifong, NIOSH, DHEW Region IV, 50 Seventh Street, N.E., Atlanta, Georgia 30323.

REGION V — Chicago
(Illinois, Indiana, Michigan, Minnesota, Ohio, Wisconsin)

Regional Program Director, Richard Kramkowski, NIOSH, DHEW Region V, 300 South Wacker Drive, Chicago, Illinois 60607.

REGION VI — Dallas
(Arkansas, Louisiana, New Mexico, Oklahoma, Texas)

Regional Program Director, George Pettigrew, NIOSH, DHEW Region VI, 1100 Commerce Street, Room 8-C-53, Dallas, Texas 75202.

REGION VII — Kansas City, Missouri
(Iowa, Kansas, Missouri, Nebraska)

Regional Program Director, Ralph Bicknell, NIOSH, DHEW Region VII, 601 East 12th Street, Kansas City, Missouri 64106.

REGION VIII — Denver
(Colorado, Montana, North Dakota, South Dakota, Utah, Wyoming)

Regional Program Director, Stanley Reno, NIOSH, DHEW Region VIII, 9th and Stout Sts. (9017 Fed. Bldg.), Denver, Colorado 80202.

REGION IX — San Francisco
(Arizona, California, Hawaii, Nevada)

Regional Program Director, Douglas Johnson, NIOSH, DHEW Region IX, 50 Fulton Street (254 FOB), San Francisco, California 94102.

REGION X — Seattle
(Alaska, Idaho, Oregon, Washington)

Regional Program Director, Walter E. Ruch, NIOSH, DHEW Region X, 1321 Second Avenue (Arcade Bldg.), Seattle, Washington 98101.

PRIVATE ASSOCIATIONS CONCERNED WITH SAFETY AND HEALTH

The following is a list of private professional organizations whose activities affect the fields of safety and health.

American Association of Industrial Nurses
79 Madison Avenue
New York, New York 10016

American Conference of Governmental Industrial Hygienists
P.O. Box 1937
Cincinnati, Ohio 45301

American Industrial Hygiene Association
66 S. Miller Road
Akron, Ohio 44313

American Insurance Alliance
85 John Street
New York, New York 10038

American Medical Association
Department of Environmental, Public, and Occupational Health
535 North Dearborn Street
Chicago, Illinois 60610

American Mutual Insurance Alliance
20 North Wacker Drive
Chicago, Illinois 60606

American National Standards Institute*
1430 Broadway
New York, New York 10018

American Society of Safety Engineers
850 Busse Highway
Park Ridge, Illinois 60068

Construction Safety Association
4750 North Sheridan Road
Chicago, Illinois 60640

Industrial Health Foundation, Incorporated
5321 Centre Avenue
Pittsburgh, Pennsylvania 15232

Industrial Medical Association
150 North Wacker Drive
Chicago, Illinois 60606

International Association of Government Labor Officials
7310 Woodward Avenue
Detroit, Michigan 48202

International Association of Industrial Accident Board and Commissions
150 Tremont Street
Boston, Massachusetts 02111

National Fire Protection Association*
60 Battery March Street
Boston, Massachusetts 02110

National Safety Council
425 North Michigan Avenue
Chicago, Illinois 60611

Unemployment Benefits Advisors, Incorporated
720 Hotel Washington
Washington, D.C. 20004

*A Congressionally-chartered organization for the development of consensus standards.

INDEX

administrative controls 40
Annual Summary 13
appeal 11, 12
asbestos 4, 83
audiogram
  annual form letters 48-50
  individual records 42
audiometric tests 41
  employee notification of results 47

Buildings and Grounds 114-118
  bonding 117
  car spots 115
  doors 115
  drains and ditches 115
  electrical and static grounding 118
  electrical equipment 116
  elevators and lifts 117
  exits 115
  file cabinet and drawers 117
  floors 116
  heating and/or air conditioning 117
  hoists, chain blocks, winches 117
  illumination 116
  lateral minimum clearances 114
  nonportable electric equipment 117
  outside walks 117
  overhead minimum clearances 114
  portable electric equipment and
    extension cords 117
  railings 116
  ramps 114
  stairs 116
  ventilation 115
  windows 115

carbon monoxide 4, 84

citations 9
  contest of 11
  posting 15
communication
  guidelines for 55
  techniques 55
community contact 53
compliance
  employer enforcement of 18
  Officer 20
  typical program 43
  voluntary 4
  with abatement requirements 11
cotton dust 4, 85

danger
  de minimus 10
  imminent 10
  nonserious 10
  serious 10

educational programs 18
employee
  request for inspection 15
  responsibility 15
  rights 15
  "shall" 9
employer
  enforcement of compliance
    responsibilities 18
  responsibilities 17
  rights 17
  "shall" 9
establishment 14

fines 9
  reduction of 10

Fire Protection Equipment 120-121
  fire alarm boxes 120
  fire doors 120
  fire extinguishers 120
  fire hydrants and hose stations
    and houses 120
  sprinkler system 120
first aid 14

hearing conservation, comments on 44
hearing conservation program 39
  continuing 42
  reasons for 46
HEW, major responsibilities 92
Housekeeping 123
how to get the job done 55

industry hiring practices 108
informal conference 11
information 53
information dissemination 51

lead 4, 87
letter of intent 37
loans 18
Log of recordable injuries and
  illnesses 12
longshoring 4
lost workdays 14

Material Handling 121
  precautions taken for 121
  storage and piling 121
meat and meat packing 4
medical treatment 14
mobile home construction 4
motivational techniques 55

National Institute of Occupational
  Safety and Health (NIOSH) 91
  develop and establish recommended
    occupational safety and
    health standards 91
  functional organization 91
  principal divisions 95
  principal duties 95
  principal responsibilities 94
  research and training 91

National Personnel
  Bureau of Labor Statistics Regional
    Offices 143
  Department of Labor 127
  National Institute for Occupational
    Safety and Health 144

OSHA Regional,
  Area and District Offices 128
Private Associations Concerned
  with Safety and Health 146
noise
  abatement program 39
  surveys 40
notification by certified mail 11

Occupational Hazard Survey Form 101
Occupational Safety and Health
  Act (1970)
  application thereof 51
  educational problems posed 108
  employee relations aspect 1, 58
  employee need to know about it 53
  goals of 58
  objective is fourfold 1
  problems 106
  requirements 1
Occupational Safety and Health
  Standards 9
  index to part 1910 63-82
Operating Equipment 121-123
  air hoses 123
  controls 122
  gas cylinders 121
  grinding wheels 122
  guards 122
  ladders 122
  layout 122
  maintenance 122
  pipe lines 121
  pressure relief valve and rupture
    discs 123
  scaffolds 123
  steam hoses 123
  temporary equipment 122
  tools 122
  unfired pressure vessels 123
  vents and flame arrestors 122
order 11
Oregon, 1972 legislation 46

penalties 9
personal protective equipment 41
Personnel 124-125
  clothing 124
  job training 125
  personal protective equipment 124
  safety meetings 125
  unsafe practices 125
protectivity 2
public relations 53
pulp and paper industry in the Pacific
  Northwest (1956) 45

"reasonable particularity" 16
record keeping 12, 94
recordable illnesses 12
recordable injuries 12
recordable occupational injuries and
   illnesses 13
reductions 18
Reference Section 114
reporting 12, 14
   failure to 14
   falsification of 14
roofing 4

safety as a separate and important
   issue 2
safety and health
   40,000 new jobs 25
Safety Equipment 118-120
   acid suits 119
   air masks 118
   bulletin board 119
   eye wash bottles 120
   gas masks 118
   goggle cleaning station 119
   life belts and ropes 119
   safety showers 118
   safety rope 119
   safety rules 119
   safety signs 119
   stretcher boxes 118
Safety Inspection Guide 111
   how to make one 111
   main points to look for 113
   unsafe conditions 113
   unsafe practices 113

safety log 35
safety management
   has changed 28
   within company structure 26
safety personnel
   Committee 31
   coordinator 26
   director 27
   qualified, competent, critically
      limited 25
Sanitation and Health 123-125
   drinking water 124
   eating 124
   lockers 124
   toilets and urinals 123
   washing and bathing facilities 123
sheet metal 4
silica 4, 88
Standards 9
   criteria for 92
   state 20
state plans evaluated 20
Supplementary Record 13

Target Health Hazards program 4, 83
target industry 4
toxicity, definition of 94
training and education 53

variances 18

wood products 4
workplace 20